Laser-based Investigation of Gas and Solid Fuel Combustion under Oxy-Fuel Atmosphere

Am Fachgebiet Maschinenbau

der Technischen Universität Darmstadt

zur Erlangung des akademischen Grades eines

Doktor-Ingenieurs (Dr.-Ing.)

eingereichte

Dissertation

vorgelegt von

Sebastian Bürkle, M.Sc., M.Sc.

aus Erbach (Odenwald)

Berichterstatter:	Prof. Dr. rer. nat. A. Dreizler
Mitberichterstatter:	Prof. Dr.-Ing. V. Scherer
Tag der Einreichung	16. Oktober 2018
Tag der mündlichen Prüfung	18. Dezember 2018

Darmstadt 2018

D17

Erklärung /Statement

Hiermit erkläre ich, dass ich die vorliegende Arbeit, abgesehen von den in ihr ausdrücklich genannten Hilfen, selbständig verfasst habe. Ich erkläre außerdem, dass ich bisher noch keinen Promotionsversuch unternommen habe.

Darmstadt, 16. Oktober 2018   Sebastian Bürkle, M.Sc., M.Sc.

Bibliografische Information der Deutschen Nationalbibliothek:
Die Deutsche Nationalbibliothek verzeichnet diese Publikation
in der Deutschen Nationalbibliografie; detaillierte bibliografische
Daten sind im Internet über dnb.dnb.de abrufbar.

© 2019 Sebastian Bürkle

Herstellung und Verlag: BoD – Books on Demand, Norderstedt

ISBN 978-3-7481-4514-1

Acknowledgements

This work originates from mytime as a research associate in the institute of Reactive Flows and Diagnostics (RSM) at the TU Darmstadt. First of all, I would like to cordially thank my supervisor and head of the institute, Prof. Dr. Andreas Dreizler, for giving me the opportunity to study the field of combustion. I very much appreciated the support he gave me with regards to any scientific or organizational questions. In addition, I am deeply grateful to Prof. Dr.-Ing. Viktor Scherer, for both his willingness to report on my thesis and his interest in my work, which are a great honor to me.

To Dr. Steven Wagner, head of the TDLAS-group, I want to express my gratitude for his personal and scientific support. His suggestions and expertise on the measurement technique have been most valuable. In this context, I want to thank the other former and current members of the group, Anna, Benni, Johannes, Gaby, Luigi, Niels, Sani, and in particular my former office-mates Felix and Olli. Without their ideas and collaborations, this work would have been impossible to achieve.

A special mention needs to be made for the members of the Collaborative Research Center/Transregio 150: "Near-Wall Reactive Flows", whose executive director I had the pleasure to be for four years now. Getting in contact with so many talented researchers of such a multitude of fields, all working together for one goal, has been quite an experience to me.

Equally, the time I scientifically spent in the CRC/Transregio 129: "Oxy-Flame" has been essential to this work. Without the collaboration with the excellent researchers of this project, my own research would have not been the same. From this large group, I would like to explicitly pick Lukas Becker and Johannes Hees for their support and the time spent during measurement campaigns at the Oxy-Fuel Burner and the Technikum. The measurements have been physically and mentally challenging, but they were worth every hour. Also, I want to pick Samim and the Sebastians - Popp and Ganter - for the valuable support and advice regarding the numerical simulations this work deeply needed.

To name a few more colleagues without mentioning their individual contributions, I want to highlight the works of Maria Angela, Angela, Christopher, Jan, Ayane, Fachgebietsmutti Wibke, the RSM-engine group, the RSM/TTD workshop, the secretaries and the soccer and running teams of RSM/EKT/STFS. My former Bachelor-student Nicole I am grateful for her outstanding work regarding the line data and for restoring my faith in both students and my skills as an advisor.

I would like to thank my friends for making the little time I had besides my research worthwhile. To explicitly name a few: Lena, the sensational seven and their girls and the AoE players. In addition, I kindly acknowledge the work of the additional proof-readers, my longtime friends Ami, Anna, Eliza and Sarah K.

Last but not the least, I would like to thank my family: my siblings and their partners for supporting me spiritually through the process of writing this thesis and my life in general. My mother and father I want to acknowledge in particular, the first for believing in my dreams and fantasies, the latter for showing me how to put them in practice. Außerdem möchte ich meinen Großeltern danken, die mir neben umfangreicher moralischer Unterstützung und Interesse an dem Thema ermöglicht haben, mein Doppelstudium durchzuführen.[1] Eva and Minor I would like to thank for the incredible support and love they gave me and for going together with me through the mentally challenging final time of writing this thesis.

It is impossible to name everyone necessary in this short paragraph, hence this list is anything but exhaustive. All of you and your support were indispensable.

Sebastian Bürkle
(Darmstadt, October 2018)

---

[1] My grandparents don't speak English. Nevertheless, I want them to fully understand this paragraph. Hence, it is intentionally kept in German in their honor.

**"If I have seen further it is by standing
on the shoulders of Giants"**

Sir Issac Newton (1642–1726)

**"I like physics,
but I love cartoons"**

Stephen W. Hawking (1942–2018)

Content

Nomenclature

Abbreviations

AR-	Anti-Reflective-
AWV	Allan-Werle Variance
BECCS	Bioenergy Carbon Capture and Storage
CARS	Coherent Anti-Stokes Raman
CCS	Carbon Capture and Storage
CCU	Carbon Capture and Use
CDF	Cumulative Distribution Function
CRN	Chemical Reactor Networks
DFB	Distributed Feedback Lasers
DL	Detection Limit
DM	Discrete Mode Lasers
FB	Foreign-broadening cell
IPCC	Intergovernmental Panel on Climate Change
LDC	Laser Diode Current Control
LES	Large Eddy Simulation
LIBS	Laser-induced breakdown spectroscopy
LIF	Laser-induced fluorescence
LIGS	Laser-induced grating spectroscopy
MIR	Mid-infrared region
NIR	Near-infrared region
OAP	Off-axis parabolic mirror
OD_e	Optical density (logarithmic to base e)
Oxy21	Atmosphere consisting of 21 $\%_{Vol.}$ O_2 and 79 $\%_{Vol.}$ CO_2
Oxy25	Atmosphere consisting of 25 $\%_{Vol.}$ O_2 and 75 $\%_{Vol.}$ CO_2
Oxy30	Atmosphere consisting of 30 $\%_{Vol.}$ O_2 and 70 $\%_{Vol.}$ CO_2
Oxy-coal	Combustion of solid fuels under carbon dioxide/oxygen atmosphere
Oxy-fuel	Combustion of fuels under carbon dioxide/oxygen atmosphere
Oxy-gas	Combustion of gaseous fuels under carbon dioxide/oxygen atmosphere
PAH	Polycyclic aromatic hydrocarbons
PD	Photodiode
ppm	Pars per million (10^{-6})
RQL	Rich Burn, Quick Mix, Lean Burn
RTD	Residence Time Distribution
SB	Self-broadening cell
SNR	Signal to noise ratio
STD	Standard Deviation
TDLAS	Tunable Diode Laser Absorption Spectroscopy
TEC	Laser Diode Temperature Control
VCSEL	Vertical Cavity Surface Emitting Diode Lasers
WHP	Wolfhard-Parker burner
WMS	Wavelength Modulation Spectroscopy

Greek notations

α	Thermal Diffusivity (Flow) or Absorption Coefficient (Spectroscopy)
γ_f	Foreign-Species Induced Collisional Broadening Coefficient
γ_s	Self-Induced Collisional Broadening Coefficient
δ_f	Foreign-Induced Pressure Shift Coefficient
δ_L	Laminar Flame Thickness
δ_s	Self-Induced Pressure Shift Coefficient
$\partial v/\partial t$	Laser Tuning
Δ	Laplace Operator
ΔH	Enthalpy Release due to Reactions
Δv	Line Width
ε	Dissipation Rate of Turbulent Kinetic Energy
ς	Stoichiometric Coefficients
η	Eddy Length Scale
η_K	Kolmogorov Length Scale
η_0	Integral Length Scale
θ	Any Scalar or Vector Quantity
$\Theta(t)$	Heaviside-Function (Residence Times) or Radiation Therm (Spetroscopy)
κ	Poisson Constant (Thermodynamics) or Eddy Wavenumber $\kappa = 2\pi/\eta$
λ	Wavelength (Optics), Mean Free Path (Continuums Mechanics) or Thermal Conductivity (Thermodynamics)
μ	Kinematic Viscosity
v	Variable in wavenumber space or absorption line central position
v_{fl}	Dynamic Viscosity
ξ	Mixture Fraction
Ξ	Chemical Activation Enthalpy
ρ	Covariance Correlation Coefficient (results) or Gas Density (Introduction)
ρ	Gas Density
σ_{AW}	Allan-Werle variance
$\sigma_{T,\,exp}$	Standard deviation of the process-induced temperature fluctuations
$\sigma_{\chi,\,exp}$	Standard deviation of the process-induced concentration fluctuations
$\sigma_{T,\,fit}$	Uncertainty of the temperature measurement
$\sigma_{\chi,\,fit}$	Uncertainty of the concentration measurement
σ_{RTD}	Variance of the RTD, distribution of residence times around the mean
τ_{ik}	Stress Tensor
τ_K	Kolmogorov Time Scale
τ_0	Integral Time Scale
ϕ	Line Shape Function (Spectroscopy) or Potential (Balance Equations)
Φ	Equivalence Ratio (Combustion) or Potential (Flow Motion)
χ	Species Mole Fraction or Scalar Dissipation Rate
χ_{lim}	Detection Limit of Species
ψ	Wave Function
$\dot{\omega}$	Source Term

Roman notations

a	Speed of Sound
A	Absorption Line Area
$A_{i,k}$	Einstein-Coefficient for Spontaneous Emission
$B_{i,k}$	Einstein-Coefficient for Absorption
$B_{k,i}$	Einstein-Coefficient for Stimulated Emission
c_p	Heat Capacity at Constant Pressure
D	Diffusion Coefficient
$E(t)$	Residence Time Distribution
E_i	Lower-state energy
E_{TKE}	Turbulent Kinetic Energy
$F(t)$	Cumulative Distribution Function
f	Frequency of Electromagnetic Wave
g	Degeneration of State
h	Enthalpy
$\hat{H}$	Hamilton Operator
I	Light Intensity
J	Angular Momentum
k	Rate Coefficients
L	Absorption length
M	Transition Moment or Molar Mass
n	Number of ideal Well-Stirred Reactors in a Cascade
OD_e	Optical density based on natural logarithm
P	Exchange Between Phases (for Mass, Momentum or Enthalpy)
$\dot{q}$	Heat Flux
$Q(T)$	Partition Sum
$\dot{Q}$	External Enthalpy Source Term
r	Reaction Rate
R	Inter-Nuclear Distance
s	Spatial Coordinate
S_{RTD}	Skewness of the residence time distribution
$S(T)$	Temperature-dependent absorption line strength
Sp	Chemical Species (e.g. CO, H_2, H_2O)
$\bar{t}$	Mean Residence Time
$\bar{t}_{PFR}$	Mean Residence Time by ideal Plug-Flow Reactor
$\bar{t}_{WSR}$	Mean Residence Time by ideal Well-Stirred Reactor
T	Temperature
$Tr(t)$	Transmissivity Therm
u_k	Velocity into Spatial Coordinates k
U	Characteristic Velocity
V	Volume or Potential (Spectroscopy)
$\dot{V}$	Volume Flow Rate
w	Mass Proportion of Elements in Species
w_{ij}	Mas Proportion of Element i in Species j
$W(t)$	Washout Function
x_k	Spatial Coordinate into direction k
Y	Species Mass Fraction
Z	Element Mass Fraction

Dimensionless numbers

Bo	Bodenstein	$Bo = uL/D_{ax}$
Da	Damköhler	$Da = \eta_0 s_L/\delta_L u'$
Ka	Karlowitz	$Ka = \eta_0/\delta_0$
Kn	Knudsen	$Kn = \lambda/L$
Le	Lewis	$Le = \alpha/D = \lambda/\rho D_i c_p$
Ma	Mach	$Ma = |u_i|/a$
Nu	Nusselt	$Nu = \alpha L/\lambda$
Pe	Peclet	$Pe = LU/\alpha$
Pr	Prandelt	$Pr = \mu/\rho\alpha$
Re	Reynolds	$Re = \rho UL/\mu$
Re_T	Turbulent Reynolds	$Re_T = \rho u'L/\mu$
Sc	Schmitt	$Sc = \mu/\rho D$

Universal constants

c	$2.997\ 924\ 58\ 10^8$ m/s	Speed of Light
e	$1.602\ 176\ 6208(98) \cdot 10^{-19}$ As	Elementary Charge
ε_0	$8.854\ 187\ 817 \cdot 10^{-12}$ As/Vm	Vacuum Permittivity
h	$6.626\ 070\ 040 \cdot 10^{-34}$ m^{2}kg / s	Planck Constant
$\hbar$		$h/2\pi$
k_B	$1.380\ 648\ 52(79) \cdot 10^{-23}$	Boltzmann-Coefficient
m_e	$9.109\ 383\ 56(11) \cdot 10^{-31}$	Electron mass
R	$8.3144598(48)$ kg m^2/(s^2 K mol)	Universal Gas Constant

Abstract

Oxy-fuel combustion has the potential to reduce the CO_2-emissions of fossil fuel power plants by burning gaseous or solid fuels under an atmosphere of carbon dioxide and oxygen. The carbon dioxide content of the combustion flue gas can easily be sequestrated from other combustion products through condensation of water vapor, liquefied and stored underground to prevent a greenhouse effect in the atmosphere. The combustion under oxy-fuel operating conditions, however, is accompanied by major changes in the combustion behavior. The underlying chemical and physical processes are complex and highly coupled, which impedes investigations and modeling.

Since tactile and most of the optical measurement techniques fail under the sensitive and simultaneously harsh environments of oxy-fuel combustion, an optical in-situ measurement system based on tunable diode laser absorption spectroscopy is developed. This system allows to investigate the thermochemical state of combustion gases with respect to the quantitative concentrations of H_2O, CO_2, O_2, CO, CH_4, HCl, OH and C_2H_2 as well as the temperatures of H_2O and CO_2. It is validated against a well-known reference flame. In combination with a newly developed and applied measurement strategy, the system even allows for a measurement of the gas residence time distribution. The measurement system relies on highly accurate absorption line coefficients for reliable measurements at low uncertainties. To fill gaps in the availability of these coefficients, the pressure broadening and shift of multiple absorption lines, caused by various combustion-relevant species, are experimentally determined.

The measurement system is applied to three oxy-fuel combustion systems in four combustion modes. First, the laminar, non-premixed methane combustion under oxy-fuel atmosphere is studied. The gas temperature profiles and concentrations of water vapor, CH_4 and CO are compared to air-blown combustion. OH concentration profiles reveal changes of the flame position. The turbulent, premixed combustion of the same fuel under air and two oxy-fuel atmospheres is studied in a 20 kW_{th} swirled combustor. Measurements of the residence time distribution of fluids in the combustion chamber provide insights into mixing and transport properties of the flow. The fuel burnout rate and zones of incomplete mixing among different flows are assessed through measurements of the flue gas thermochemical state. A comparison to equilibrium calculations allows conclusions regarding the reaction progress.

In the same combustor, the co-firing of three different solid fuels in an assisting gas flame is investigated for a combined thermal power up to 40 kW_{th}. Measurements of the gas temperature and quantitative concentrations of three combustion gases in combination with equilibrium calculations allow to estimate the burnout rates of volatiles and char. The pure combustion of torrefied biomass under air and oxy-fuel combustion atmosphere is investigated in a 60 kW_{th} close-to-application facility. Similarities between different flames as well as parallelisms to the other experiments are evident.

Overall, these experiments demonstrate and quantify various changes and similarities in the behavior of gaseous and solid fuel combustion under air compared to oxy-fuel atmosphere. Hence, this work provides an improved understanding and broader experimental background of the oxy-fuel combustion.

Kurzzusammenfassung

Die Verbrennung gasförmiger oder fester Energieträger unter Oxy-Fuel-Atmosphäre, einem Gemisch aus Kohlenstoffdioxid und Sauerstoff, birgt ein großes Potential die CO_2-Emissionen von Kraftwerken zu reduzieren. Der Kohlenstoffdioxidanteil der Abgase dieses Prozesses lässt sich leicht von Wasserdampf, dem anderen Hauptreaktionsprodukt dieses Verfahrens, durch Kondensation des Dampfes abscheiden. In nachgelagerten Prozessen kann das CO_2 dann verflüssigt und in unterirdischen Gesteinsstrukturen eingelagert werden, wodurch ein Beitrag zum anthropogenen Treibhauseffekt verhindert wird. Das Verbrennungsverhalten der Energieträger wird allerdings durch die veränderte Verbrennungsatmosphäre entscheidend beeinflusst. Die diesen Veränderungen zugrundeliegenden chemischen und physikalischen Mechanismen sind hochkomplex und in vielerlei Hinsicht gekoppelt, was eine Modellierung erschwert und experimentelle sowie numerische Untersuchungen behindert.

In dieser Arbeit wird ein Messsystem entwickelt das Untersuchungen der Oxy-Fuel-Verbrennung ermöglicht. Da aufgrund der anspruchsvollen und zugleich sensitiven Zustände innerhalb der Verbrennungsumgebung sowohl taktile als auch der Großteil an optischen Messverfahren ausscheiden, wird ein optisches Messsystem auf Basis der Absorptionsspektroskopie mittels durchstimmbarer Diodenlaser realisiert. Dieses ist in der Lage, den thermochemischen Zustand der Verbrennungsgase berührungslos zu erfassen. Im Speziellen werden die quantitativen Konzentrationen der Verbrennungsspezies H_2O, CO_2, O_2, CO, CH_4, HCl, OH und C_2H_2 sowie die Gastemperaturen von H_2O und CO_2 gemessen. Validiert wird das System hinsichtlich der thermochemischen Messungen anhand einer gut untersuchten Referenzflamme. In Kombination mit einer neuentwickelten Messprozedur erlaubt das System überdies hinaus eine Erfassung der Verweilzeitverteilung gasförmiger Verbrennungsbestandteile. Für eine hochgenaue Bestimmung der quantitativen Prozessparameter benötigt das Messsystem zuverlässige Liniendaten mit geringen Messunsicherheiten. Da innerhalb dieser Daten große Kenntnislücken vorhanden sind, werden die durch verschiedene verbrennungsrelevante Moleküle hervorgerufenen Druckverbreiterungen und -verschiebungen ausgewählter Absorptionslinien experimentell bestimmt.

Das Messsystem wird an drei Oxy-Fuel-Verbrennungssystemen in vier unterschiedlichen Verbrennungsarten angewendet. Zunächst wird eine laminare, nichtvorgemischte Methanflamme unter Oxy-Fuel-Atmosphäre untersucht. Die Gastemperatur und Konzentrationen an Wasserdampf, Methan und Kohlenstoffmonoxid in der Verbrennung werden mit denen innerhalb einer luftbasierten Verbrennung verglichen. Veränderungen in der Flammenposition werden anhand von OH-Konzentrationen detektiert. Des Weiteren wird die turbulente, vorgemischte Verbrennung von Methan unter Luft und zweier Oxy-Fuel-Atmosphären unterschiedlichen Sauerstoffgehaltes an einer 20 kW_{th}-Anlage untersucht. Die Ausbrandrate des Kraftstoffs sowie Zonen unvollständiger Durchmischung werden anhand des thermochemischen Zustandes bestimmt. Durch einen Vergleich des Zustandes mit Gleichgewichtssimulationen können Rückschlüsse bezüglich des Reaktionsfortschritts gezogen werden. Messungen der Verweilzeitverteilung von Verbrennungsgasen innerhalb der Brennkammer ermöglichen darüber hinaus Einblicke in Transporteigenschaften der Strömung.

In der gleichen Brennkammer wird die Verbrennung dreier fester Energieträger in Oxy-Fuel-Umgebungen mit Hilfe einer unterstützenden Gasflamme bei kombinierten Leistungen von bis zu 40 kW_{th} untersucht. Messungen der Gastemperatur und der Hauptspezies ermöglichen es durch einen Vergleich mit Gleichgewichtssimulationen, den Ausbrandgrad der Volatilen und des Koks abzuschätzen. Die reine Verbrennung von torrefizierter Biomasse unter Luft und Oxy-Fuel-Atmosphären wird in einer anwendungsnahen 60 kW_{th}-Anlage untersucht. Ähnlichkeiten zwischen verschiedenen Flammen und Parallelitäten zu den vorhergehenden Experimenten werden aufgezeigt.

Insgesamt ist es innerhalb dieser Untersuchungen gelungen, verschiedene Änderungen und Parallelitäten des Verbrennungsverhaltens gasförmiger und fester Energieträger in Oxy-Fuel-Atmosphäre im Vergleich zu Luft zu demonstrieren und zu quantifizieren. Daher trägt diese Arbeit zu einem verbesserten Verständnis und einer umfassenderen experimentellen Datenlage der Oxy-Fuel-Verbrennung bei.

List of Figures

List of Tables

1 Introduction

Since the upper Paleolithic (late stone age) 50 000 – 10 000 years ago humans have utilized fire for heating and illumination. Some researchers even believe that without the invention of cooking over solid-fuel fires and the accompanied increase of nutrients in the diet, a reduction of the gut size and a consecutively allowed increased brain size of the homo erectus would not have been possible [1]. Since the beginning of the industrial age in the 18th century combustion processes are used to convert the primary energy of gaseous and solid fuels into other forms of energy, such as mechanical or electrical, on large scales. Approximately 80% of the global primary energy demand is nowadays satisfied by the conversion of fossil fuels, which can be subdivided into a contribution of solid fuels by approximately 29 % and of gaseous fuels by 21 % (liquid fuels 30 %, statistics from 2014, International Energy Agency [2]). The downside of the combustion benefits have always been the emissions of pollutants. While state-of-the-art systems for reduced emissions of sulfur oxide, hydrogen chloride, carbon monoxide, lead and particulate matter (e.g. dust or soot) are nowadays established technologies, a more recent focus of research is on reducing nitric oxide and carbon dioxide (CO_2) emission. The latter emissions affect life-forms not only on a local, but on a global scale.

A multitude of evidence was found on CO_2 emissions causing global warming [3], which is expected among other effects to lead to rising seawater levels, reduced biodiversity, droughts and floods, regional heat records and wildfires. To limit the effects of global warming but at the same time to allow economic growth, 195 nations agreed in the 2015 Paris agreement [4] to keep the increase in the global average temperature well below 2 °C above pre-industrial levels. Furthermore, it was decided to preferably limit the increase to 1.5 °C, since this would substantially reduce the risks and effects of climate change. As a consequence, many industrialized countries which contribute to a large share of the global CO_2 emissions have decided to cut back on combustion of fossil fuels. However, even though the contribution of renewable energies to the power generation in these countries is on the rise, the combustion processes will continue to give an important contribution to their energy production within the coming decades.

Moreover, plenty of developing and emerging countries or geopolitical regions, such as South Africa, Southeast Asia, India and China, but also some industrialized countries such as the USA, insisted on a necessity of economic growth and prosperity for their countries. As the power generation through fossil fuels is comparatively cheap, these countries decided to retain or even increase the current levels of fossil fuel combustion. As the latter three countries alone contributed to approximately half of the worldwide CO_2 emissions in 2015, the continued combustion of fossil fuels, in particular of coal, in these countries pose a non-negligible thread to the goals of the Paris agreement.

One way to overcome these conflicts of objectives between cheap economic growth and reduced CO_2 emissions, besides increased power plant efficiencies, is the utilization of *carbon capture and storage* (CCS) technologies [5]. In these processes, the carbon dioxide from fossil fuel power plants is captured, liquefied, transported to a storage site (usually underground geological formations such as deep saline formations or former gas and oil fields) and deposited in environments sealed from the atmosphere. The aim is to prevent the release of large quantities of CO_2 into the atmosphere by storing it under liquefied or supercritical conditions. In addition to CCS, a sequestration of flue-gas CO_2 allows for a utilization in chemical processes (*carbon capture and utilization,* CCU), such as urea-production or generation of carbon-neutral renewable fuels. A combination of the CCS techniques with the combustion of bioenergetics fuels, such as biomass, even allows for negative CO_2 emissions as CO_2 is effectively removed from the atmosphere (*bioenergy carbon capture and storage,* BECCS). According to the Intergovernmental Panel on Climate Change (IPCC [3]) many emission forecast scenarios and models could not limit warming to below 2°C without a significant utilization of bioenergy, CCS and their combination (BECCS). Compared to competing CO_2 emission free technologies, such as renewable energies, the migration costs are significantly lower for CCS due to a possible retrofitting of existing power plants. The IPCC estimates the economic potential of CCS to be between 10% and 55% of the total carbon mitigation effort until the year 2100.

Due to a low public acceptance, CCS is not likely to be implemented on large scales within European countries. However, the aforementioned countries with prolonged or increased utilization of fossil fuel combustion are required to use CCS, as otherwise the achievements of other countries are jeopardized. China and the USA are planning to increase their efforts in CCS research and implementation. As an example, the US National Academy of Engineering lists the development of carbon sequestration methods as one of the 14 grand challenges for engineering in the 21st century. More than half of the planned or operating large-scale CCS facilities are located in either of the two countries [6]. By 2040, the IPCC expects 70 % of the electricity still generated from coal to come from power plants equipped with CCS, of which ~50 % are expected to be located in China.

Various methods have been proposed for carbon capture, which can be classified into the following:

- Pre-combustion capture,

- Post-combustion capture,

- Oxy-fuel combustion.

In pre-combustion capture the hydrocarbon fuels are converted to obtain H_2 and CO through gasification or partial oxidation [5]. Using H_2O steam reforming, the resulting syngas itself is converted to CO_2 and H_2 through the water-gas shift reaction. While the CO_2 content is then separated and stored, the H_2 fraction is used for emission-free power generation. However, this technique requires either gas or steam turbines based on hydrogen-combustion or, alternatively, fuel cells. All of these techniques are under development but comparatively challenging and expensive and are not eligible for retrofitting. In post-combustion capture, a scrubber (e.g. chemical sorbents or membrane systems) separates the CO_2 content off the flue gas. Even though this technique involves minimum changes to existing power plants, the overall effort and decrease in plant efficiency is high [7] (estimated to 10–14 % loss in overall efficiency), as the CO_2 concentrations in the flue gas are low (~3–15 %$_{Vol.}$) and the separation systems therefore complex, expensive and energy-intensive.

In oxy-fuel combustion, the fuel is burned in a nitrogen-free atmosphere instead of air. The technique was originally proposed for enhanced oil recovery (gas-injection into subterranean oil fields) by Abraham et al. [8]. Oxy-fuel combustion in oxygen-enriched environments have been extensively used in processes which require higher adiabatic flame temperatures, such as glass furnaces, boilers or metal cutting processes. To limit the resulting flame temperatures, a fraction of the flue-gas is cooled and recirculated into the combustion chamber oxygen flow, which is provided by an air separation unit. The flue gas of recirculated oxy-fuel combustion is mainly composed of CO_2 and water vapor (H_2O), which can be easily separated through condensation. The combustion atmosphere of this technique, however, consists of a mixture of mainly oxygen and CO_2. Oxy-fuel combustion allows for a retrofit of older power plants and promises lower migration costs and higher plant efficiencies as the other techniques. Additional benefits of the technique are reduced emissions of nitrogen oxides (NO_x), as the oxidizer contains little to no nitrogen and only fuel-bound nitrogen remains [9]. Furthermore, oxy-fuel combustion allows a direct influence on the flame temperature by variation of the oxygen content, which allows greater freedom in the burner design.

However, the combustion of solid fuels under oxy-fuel atmosphere is accompanied by major changes in the combustion behavior, which will be discussed in detail in subsection 2.1.5. To give a few examples, due to the higher molar heat capacity of CO_2 compared to N_2, the adiabatic flame temperatures and flame speeds for identical oxygen contents are lower, which possibly influences flame stability. The diffusivity of oxygen in nitrogen exceeds that of oxygen in carbon dioxide by ≈ 1.43, which causes a need for higher oxygen-contents in oxy-fuel combustion. As the flue gasses are recycled, higher concentrations of sulfur and nitrogen oxides as well as water vapor have to be expected in the fresh gasses, which influences the combustion chemistry. At high temperatures, a fraction of the CO_2 converts into CO during the combustion process due to several chemical reactions.

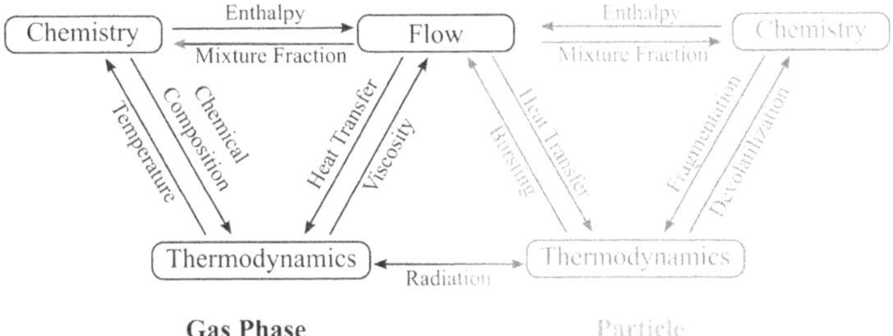

Fig. 1.1: Examples of mutual dependences between chemistry, flow and thermodynamic properties in the combustion of gaseous or solid fuels.

To improve the efficiency of oxy-fuel combustion processes and reduce associated emissions of pollutants and on the other hand allow predictions of the combustion behavior through improved models and simulations, a detailed analysis of the underlying processes is crucial. The processes involved in combustion in general, but in particular under oxy-fuel combustion atmosphere, are complex, as various processes, time-and length scales are simultaneously involved and strongly interlinked.

Fig. 1.1 depicts these mutual dependencies between the gas phase and solid fuel combustion processes in a simplified diagram. The combustion processes are influenced by the gas phase flow field as well as by the chemistry and thermodynamic properties of both the gas and the particle. Each of these disciplines cover various regimes, length- and time-scales and require knowledge of elementary processes, balances and state equations. In addition, heat transfer by radiation and walls play a crucial role in gaseous and solid fuel flames and mutually affect the chemistry, flow and thermodynamics. For example, radiation affects the particle thermodynamics by increasing the heat-up rate, whereas the walls influence the gas-phase chemistry and thermodynamics through catalytic effects and enthalpy-losses. Most practical gas or solid fuel combustion applications cannot be described by one or only a few of these processes, as they are highly interlinked and therefore inseparable.

As investigations supporting our fundamental understanding are sparse, following a stepwise approach from generic to close-to-application combustion systems by steadily increasing the complexity of the investigated processes has proven beneficial.

1.1 State of research

The following section gives an overview on the state of research of the gaseous (oxy-gas combustion) and solid fuels (oxy-coal combustion[2]) under oxy-fuel combustion atmosphere, largely following the stepwise approach from generic experiments to close-to-application combustion systems. Even though both chemical conversions occur under oxy-fuel combustion atmospheres, their fundamental processes and practical realizations of combustion systems largely differ. Hence, their state of research will be discussed separately. The different measurement techniques allowing investigations of oxy-fuel combustion processes will be reviewed separately at the beginning of chapter 3.

---

[2] For simplicity, this work utilizes the term "oxy-coal combustion", even though other types of solid fuels, such as biomass, are investigated besides coal as well.

1.1.1 Oxy-Gas combustion

While application-relevant processes are preferably study through natural-gas flames, generic experiments usually rely on less complex fuels. To investigate isolated effects of oxy-gas combustion, laboratory-scale methane flames are favorably studied, as the underlying chemical reactions are well known and easier to describe.

Various measurements on single quantities of oxy-gas combustion at small-scale, generic burners can be found in literature, such as the radiative heat flux and flame height [10], flame speed [11–13], flame thickness [14] and soot formation [15]. For the investigations, methane was burned in laminar, premixed Bunsen [11,14], laminar flat-flame [15], laminar premixed ignition-chamber [12] and laminar non-premixed jet [10] burners. The experiments show large differences to the respective quantities under air-blown combustion, which are attributed to the different properties of CO_2 compared to nitrogen. To achieve similar combustion properties in terms of adiabatic flame temperature, flame speed and flame thickness for oxy-gas combustion compared to air-blown gas combustion, the studies suggest a higher necessary oxygen-content in the combustion atmosphere. Depending on the quantity under investigation, the proposed or measured optimal content in order to achieve similar properties ranges from 25–35 %$_{Vol}$. Due to the higher CO_2 content, oxy-fuel flames were observed to have a significantly higher radiative heat transfer than comparable air flames [16]. While the thermochemical state, i.e. the temperature and the species concentrations, of the laminar, premixed combustion has already been studied using Raman/Rayleigh spectroscopy [17], detailed investigations of generic laminar, non-premixed oxy-fuel flames with well-known boundaries are still sparse.

More complex phenomena are favorably studied in turbulent swirl [18–21] or jet flames [22]. In these publications, the flame temperature and mixture fraction [18], flame structure [18,20], thermochemical state [17,20,22], emission characteristics [19] and flow field [20,21,23] of methane/oxy-fuel combustion is studied using Raman/Rayleigh spectroscopy, gas analyzers, OH*-chemiluminescence imaging and particle image velocimetry, respectively. Kapadia [20] showed that the flame properties of Oxy30 (30 %$_{Vol}$ O_2/70 %$_{Vol}$ CO_2) in a swirled gas-turbine burner are most similar to air-blown combustion. Below Oxy25, flame quenching can be expected. The authors found similarities between the flame variations caused by different oxygen-levels in the oxy-fuel combustion atmosphere and those induced by different equivalence ratios of air-blown combustion. Sevault [17] found increased CO levels during combustion of oxy-fuel flames compared to the air-blown situation in non-premixed jet flames. The thermochemical state of premixed, swirled oxy-fuel/methane flames in combination with the flow-field, however, has not been investigated yet.

Examples of larger-scale oxy-gas experiments were performed by Kim et al. [24] and Becher et al. [25] regarding NO reduction and proof-of concept controlled staging in 0.2 MW_{th} oxy-methane burners. Kapadia [20] investigated the flow field and flame shape of a pressurized, oxy-methane swirl burner of up to 0.15 MW_{th}. Tan et al. [26] investigated the flue gas composition of a 0.3 MW_{th} down firing facility burning natural gas in oxy-fuel combustion atmosphere in a flue gas recirculation through ex-situ gas analysis. The authors found reduced NO_x emissions and an improved plant efficiency due to lower gas volumes and better operational flexibility for the oxy-fuel process compared to air-blown operation.

At higher thermal powers, investigations are sparse and mainly focus on power-plant efficiency, techno-economic considerations and thermodynamics based on plant simulations rather than detailed investigations of sub-components or effects [27–30]. Bolland et al. [29,30] studied the design of oxy-fuel gas turbines fired with natural gas as a CO_2 removal option and reported a possibility of a 90% reduction in the atmospheric discharge of carbon dioxide.

1.1.2 Oxy-Coal Combustion

Various literature is available on pure [31–36] or gas-assisted [33,37–40] oxy-coal combustion, wherefore an overview on relevant literature and investigations will be given. In terms of generic solid fuel particle combustion, the investigations span from single particle to particle-cluster combustion. In single particle combustion, which has been investigated by injecting isolated coal particles into the flue gas of well controlled laminar flow reactors, different stages of the combustion process can be separated and studied step-by-step, such as devolatilization and formation of volatiles around the particle [41,42], ignition and combustion of volatiles [33,39,40,43–45], and char combustion [41–43,45–47]. Bai et al. [43] performed experiments in a optically accessible drop-tube furnace using high-speed imaging. The authors reported volatile and char combustion of large coal particle (150–212 µm) to take place sequentially but for small particles (106–150 µm), however, to occur simultaneously. Shaddix and Molina [40] showed that an oxy-fuel combustion atmosphere retards the onset of ignition and increases the duration of devolatilization. This effect is explained by the higher molar specific heat of oxy-fuel and its tendency to reduce the local radical pool. A higher oxygen concentration in the oxy-fuel was demonstrated to lead to shorter ignition delays and devolatilization times. Similarly to findings of oxy-gas combustion, they reported the combustion properties (particle ignition and devolatilization) of an Oxy30 combustion atmosphere in [39] to lead to similar results as air-blown combustion.

In particle-cluster combustion, the volatile clouds of particles overlap and impede a description as isolated particles [48,49]. In realistic combustion chambers with varying particle densities usually a combination of both regimes, single and particle-cluster combustion, occurs. Investigations primarily focused on measuring flow fields, flame temperatures and structures, species concentrations and particle as well as flue gas temperatures. The power of burners under investigations range from single particle combustion [33,39–41], <1 kW$_{th}$ [50], 2–3 kW$_{th}$ [15,33], 20–60 kW$_{th}$ [18,31,32,37,38,51], 200–800 kW$_{th}$ [34,52] up to 30 MW$_{th}$ [35] in the famous test-facility "Schwarze Pumpe".

However, in the intermediate range of 10–100 kW thermal power, detailed and comprehensive experimental and numerical investigations of the oxy-fuel combustion are still sparse. In particular, investigations of multiple quantities, such as flow field, flame structure and temperature, flue gas composition and temperature in similar combustion systems in terms of thermal power, geometry, fuel and oxygen content are rare. Balusamy et al. [37,38] measured the flow field, soot distribution and reaction zone of a 20 kW$_{th}$ gas-assisted oxy-coal swirl burner using laser Doppler velocimetry, particle image velocimetry (PIV), laser induced fluorescence (LIF) of the OH-radical and laser-induced incandescence. Hees et al. [31,32] and colleagues of the same work group [53,54] measured the flame structure in a 60 kW$_{th}$ coal burner under air and oxy-fuel atmosphere using chemiluminescence and species concentrations within the reaction zone and the flue gas using ex-situ gas-analyses. They found the in-flame CO concentrations during oxy-fuel combustion to exceed those generated by air-blown combustion up to tenfold.

For more detailed information about oxy-coal combustion research in general, the interested reader is referred to two excellent review articles by Toftegaard et al. [7] and Scheffknecht et al. [55].

1.2 Aims and Structure of this Work

In this work, the combustion of gaseous and solid fuels under oxy-fuel combustion atmosphere is compared to conventional air-blown combustion. Experimental results of the thermochemical state and residence time distribution are compared to generic models and calculations for an identification of relevant effects from the multitude of relevant processes (fig. 1.1). In addition, the quantitative results of this work may serve as validation data for future, more elaborated numerical simulations. Following the strict approach of a stepwise increase of complexity, this work investigates similar quantities at different types of combustion processes ranging from generic small-scale flames to large-scale, close-to-application combustors. These overall goals allow to subdivide the aims of this work:

- to develop a measurement system suitable for measuring multiple quantities in oxy-fuel combustion systems in-situ in a wide range of applications,

- to validate the measurement system against a well-investigated combustion system,

- to investigate oxy-gas combustion in a laminar, non-premixed, generic setup which simulates the oxy-fuel combustion of volatiles around solid fuels with few relevant processes present,

- to investigate turbulent, premixed oxy-gas combustion on a larger, intermediate scale,

- using the same test-rig, to investigate a combined oxy-gas/oxy-coal combustion process, but with different processes relevant than in large-scale applications,

- to investigate a pure oxy-coal combustion process on a large scale, close-to application test-rig with multiple processes relevant.

From these aims, the structure of this work is deducted as following:

Chapter 2 introduces the fundamental concepts and techniques necessary to understand the results of this work. The chapter is subdivided into three sections. In the first, elementary models of gas and solid fuel combustion are provided together with necessary tools for a mathematical description. Important changes when switching from air-blown to oxy-fuel combustion are outlined. In the second subsection, the utilized measurement technique for an investigation of the thermochemical state is introduced together with the necessary quantum-mechanical principles. The chapter concludes with an introduction to possible measurement principles allowing investigations of the residence time distribution, which is the distribution of durations fluid elements spend within the combustion chamber.

Chapter 3 focuses on the practical realization of the optical measurement technique for an investigation of the thermochemical state. The measurement system had to be developed specifically for the applications of this work and, due to its novelty, itself poses a scientific progress. Hence, the features and limitations of the measurement system are discussed in detail. In addition, measurements of fundamental molecular parameters were necessary for a detailed understanding and improvement of the measurement system, which are outlined at the end of this chapter.

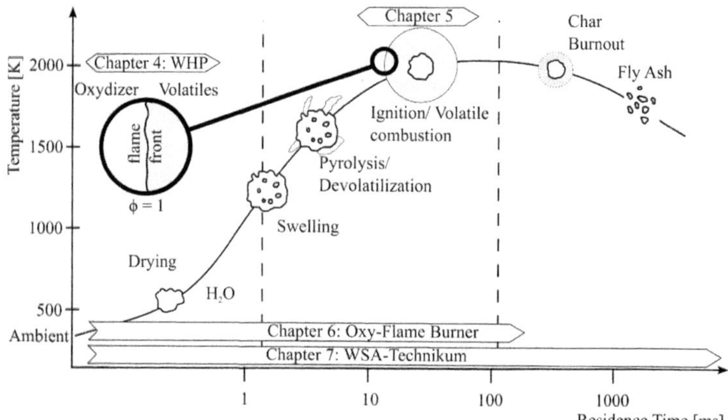

Fig. 1.2: Schematic of the combustion process in pulverized solid fuel firing; location and structure of this work.

The investigations of gas and solid fuel combustion under oxy-fuel atmosphere and the corresponding structure of this work can be illustrated best by fig. 1.2 which depicts the different stages of solid, pulverized fuel combustion and will be described in further detail in subsection 2.1.4. Starting from drying in a hot environment, the particles start to swell and melt, which leads to a release of volatile matter. These volatiles issue a gaseous fuel atmosphere around the particles which, in combination with the surrounding oxidizer, are ignited and burned. After these homogeneous reactions are completed, the heterogeneous char burnout of the remaining particle occurs. After this process is finalized, only ash and flue gasses remain of the particles.

This schematic (fig. 1.2) allows to locate the different segments of oxy-fuel related research in this work:

In **Chapter 4**, the homogeneous combustion of volatiles around a coal particle is abstracted by a laminar, non-premixed flame (Wolfhard-Parker Burner). A well-investigated air-blown CH_4 flame allows a validation of essential components of the measurement system developed in the previous section. Subsequently, the combustion of a methane flame under oxy-fuel combustion atmosphere is investigated at the same burner and relevant changes are discussed.

In **Chapter 5**, the combustion of volatiles is abstracted on a larger scale through a combustion of methane in a turbulent, non-premixed burner. The thermochemical state is measured and compared to equilibrium calculations of the flue gas. Differences between combustion under air and two different oxy-fuel atmospheres with different oxygen content are outlined. In addition, a measurement technique for an acquisition of the residence time distribution is developed and evaluated. Using results of measurements with this technique, conclusions regarding the flow field in the combustor are drawn and compared to results of the thermochemical investigations.

In **Chapter 6**, the previously investigated methane flames serve as a stabilization flame for turbulent, gas-assisted solid fuel flames in the three combustion atmospheres. The flue gas thermochemical state of three different solid fuels (Rhenish lignite, Prosper Haniel bituminous coal, torrefied biomass) is compared to equilibrium calculations, which indicate that a high reaction progress of the solid fuel particles towards volatile combustion is reached. However, it is shown that the flames are incapable of a complete char burnout. Similar to the previous chapter, measurements of the residence time distribution allow conclusions regarding the flow field.

In **Chapter 7**, the incomplete char burnout is overcome in a close-to-application facility, which allows to fire pure solid-fuel flames under turbulent air and oxy-fuel environments. One of the solid fuels of the previous chapter, torrefied biomass, is burned under air and two oxy-fuel atmospheres. Under air, the local equivalence ratio is altered. Measurements of the thermochemical state at various downstream positions allow to study the flame structure and reactions.

Chapter 8 sums up this work and gives an outlook on future developments.

1.3 Scientific Contribution

Anticipating some results of the subsequent chapters, this work contributes to two fields of research: optical measurement systems for combustion research and investigations of oxy-fuel combustion systems. The latter in turn can be subdivided into the fields of homogeneous gas-phase combustion and heterogeneous solid-fuel combustion.

To the field of optical measurement systems this work contributes in first place by developing an unparalleled in-situ measurement system for quantitative concentrations of eight combustion-relevant species and gas-phase temperatures of two species, which is fast, quasi-simultaneous, application-dependently scalable, contact-free and suitable for harsh environments. In contrast to a vast majority of comparable systems, this measurement system is validated under relevant conditions at a well-investigated combustion system. Secondly, this measurement system required knowledge of fundamental absorption line coefficients. The experiments of this work let to a unique coefficient-dataset of high accuracy, with the majority of the coefficients reported for the first time.

In third place, as measurement techniques for obtaining the residence time distribution in combustion systems reported in literature are either not applicable to realistic combustor designs or provide little meaningful results, a novel measurement strategy was developed and evaluated in this work. This strategy is shown to deliver highly-accurate measurements.

To the field of oxy-fuel gas-phase combustion this work contributes by investigating a wide range of different combustion systems in terms of size, turbulence, thermal power and complexity with the same measurement system and thus allowing to draw similarities between the different systems and relevant effects. From this contribution, the investigation of the thermochemical state of a non-premixed, laminar oxy-fuel/methane co-flow flame can be named in the first place. The results of these investigations support future numerical simulations by serving as a validation system. In second place, the careful analysis of differences between turbulent methane combustion in air and different oxy-fuel atmospheres in terms of residence time distribution and flue-gas thermochemical state can be named. These studies let to a detection of an incomplete reaction progress due to quenching in the flames, which ceased for high O_2 contents in the oxy-fuel combustion atmosphere.

The co-firing of solid fuels in the methane flame let to a third contribution, as the burnout of volatiles and char could be clarified in dependence on the respective fuel and the combustion atmosphere by comparing the flue-gas species concentrations to equilibrium calculations. Three different solid fuels were investigated which included representatives of the most-widely utilized fossil fuels lignite and bituminous coal and a promising regenerative alternative for these coals.

The regenerative alternative, torrefied biomass, was subject to a fourth contribution. By measuring the thermochemical state of this fuel's combustion under air with an alteration of the near-burner equivalence ratio and two oxy-fuel atmospheres in a close-to-application facility, the practicability of this fuel as an alternative to fossil coals was studied.

2 Fundamentals

This chapter outlines the fundamentals of the two pillars this work is based upon: the oxy-fuel combustion and the experimental methods for its research. The first part includes a brief description of the governing equations necessary to describe the processes involved, an introduction to turbulence, combustion of gaseous and solid fuels, as well as brief considerations on oxy-fuel combustion.

In the second part the basics of absorption spectroscopy are depicted, with a special focus on tunable diode laser absorption spectroscopy (TDLAS). This method is the primary measurement technique used for the investigations in this work. Finally, fundamental techniques for the investigation of the flow field in terms of residence time distribution are discussed and an overview of the most important literature in this field is given.

The primary aims of this chapter are:

- to introduce the basic terms of turbulence, combustion of gaseous and solid fuels and oxy-fuel combustion,

- to provide a background of the fundamentals and characteristics of TDLAS and

- to depict the canonical concepts of residence time measurements and evaluate their current utilization in combustion research.

As the scope of this work only allows to cover the most important aspects of each field, the interested reader is referred to related literature in each section.

2.1 Basic Concepts of Flow and Combustion

While the fundamentals of conventional gas combustion are largely understood and the processes involved can nowadays be described and modeled with satisfactory results, the scientific advance of solid fuel combustion is still sparse. Due to the complex structure of the fuel, the large numbers of chemical reactions involved and the broad range of involved time and length scales, a comprehensive description through models and numerical simulation tools is still unsatisfactory. Hence, the scientific and engineering base of oxy-fuel combustion is fragmentary and far from complete.

This section describes the characteristics and models of gas combustion and gives an overview of solid fuel combustion. A brief introduction into the features of oxy-fuel combustion and possible technical realizations are given.

2.1.1 Balance Equations

A holistic description of combustion systems requires a variety of quantities to be preserved. Most commonly, thermo-fluid mechanics model fluids by a continuous mass instead of individual molecules, which requires the mean free path of the particles to be significantly smaller than typical length scales of the system ($Kn \ll 1$)[3]. For conventional combustion systems, this prerequisite is fulfilled. The conservation of mass, which originates from the divergence theorem, is one of the fundamental concepts of non-relativistic systems. In differential form it can be stated as:

---

[3] The Knudsen-number $Kn = \lambda/L$ describes the ratio between the mean free path of molecules to typical length scales of the application.

$$\frac{\partial \rho}{\partial t} + \frac{\partial \rho u_k}{\partial x_k} = P_{PC,m} \qquad \text{(Eqn. 1)}$$

Accordingly, the temporal change of the mass within a given volume balances with the mass fluxes through the surfaces of the volume. Herein, ρ is the mass density of the fluid and u_k the three spatial components of the velocity in the directions x_k. In classical one-phase flows, the source term $P_{PC,m}$ is zero. However, in two-phase flows, phase-change occurs e.g. due to evaporation of droplets, condensation of slag on combustor walls or devolatilization of coal. In this case, $P_{PC,m}$ allows a mass-coupling between the two phases.

Similar to the conservation of mass, the conservation of momentum is one of the cornerstones of thermo-fluid mechanics and originates from the divergence theorem. The combination of both principles is commonly referred to as Navier-Stokes equations, as they were independently developed by Claude Navier [56] and George Stokes [57]. The axiom of momentum conservation is a direct consequence of Newton's laws of motion and represents their transformation to fluids. According to the equations, the momentum ρu_i is conserved and only changes due to volumetric ($\partial \phi / \partial x_i$) or external forces. In the three-dimensional, differential form it calculates to:

$$\frac{\partial(\rho u_i)}{\partial t} + \frac{\partial(\rho u_i u_k)}{\partial x_k} = \frac{\partial}{\partial x_k}\tau_{ik} - \frac{\partial p}{\partial x_i} - \rho\frac{\partial \phi}{\partial x_i} + P_{PC,I_i} \qquad \text{(Eqn. 2)}$$

The stress-tensor τ_{ik} depends on the fluid, which is normally modeled as a Newtonian fluid with a linear relationship between viscous stress and strain rate. Similarly to the conservation of mass, P_{PC,I_i} represents an exchange of momentum between the phases. p is the local pressure. The result of the Navier-Stokes equations is the flow-field, which, in most cases, has to be calculated numerically even for isothermal and non-reactive flows.

In combustion, chemical reactions and heat release are additionally present, which require a description of the fluid composition. This composition can be described by either the mass fraction Y_i or the mole fraction χ_i of each species:

$$Y_i = \frac{m_i}{\sum m_i} \text{ and } \chi_i = \frac{n_i}{\sum n_i} \qquad \text{(Eqn. 3)}$$

with the mass density m_i and the number density n_i of each species. Changes in the mass fractions are described by transport equations:

$$\frac{\partial(\rho Y_i)}{\partial t} + \frac{\partial(\rho Y_i x_k)}{\partial x_k} = -\frac{\partial}{\partial x_k}\left(\rho D_k \frac{\partial Y_i}{\partial x_k}\right) + \dot{\omega}_{i,Y} + P_{PC,Y_i} \qquad \text{(Eqn. 4)}$$

where the temporal changes of the species (first term) together with the convectional species transport (second term) are balanced by the species diffusion (third term) and the species source term $\dot{\omega}_{i,Y}$. The species diffusion is modeled according to Fick's first law, which assumes a linear relation between the diffusive transport and the species gradient. The slope of this relation is described by the diffusion coefficient D_k. The source term originates from chemical reactions in the combustion process which convert one species into others and will be described in subsection 2.1.3. Hence, the source terms and diffusive transports of all species need to balance each other out ($\sum \dot{\omega}_i = 0$ and $\sum_k D_k \partial Y_i / \partial x_k = 0$). The term P_{PC,Y_i} allows a conversion of species between the phases, e.g. through devolatilization of solid fuels or adsorption in catalysts.

In addition, the chemical reactions release heat into the flow, which leads to non-isothermal conditions. The corresponding conservation of energy is often converted into a conservation of enthalpy h.

Assuming low Mach-numbers[4] Ma<<1 and only gravitational volume forces, which are both valid in conventional combustion systems, the enthalpy balance reads as:

$$\frac{\partial(\rho h)}{\partial t} + \frac{\partial(\rho h u_k)}{\partial x_k} = -\frac{\partial q_k}{\partial x_k} + \dot{Q} + P_{PC,h} + \dot{\omega}_h \qquad \text{(Eqn. 5)}$$

$\dot{Q}$ denotes the external enthalpy source or sink term, which, in the context of combustion, is dominated by radiative heat transfer.[5] In most oxy-fuel applications this term cannot be neglected due to the enhanced radiative heat transfer by CO_2. $P_{PC,h}$ describes the exchange of enthalpy between the phases, e.g. between solid particles and the gas phase, and $\dot{\omega}_h$ the enthalpy source term due to chemical reactions. The local heat flux density q_i calculates to:

$$q_i = -\frac{\lambda}{c_p}\frac{\partial h}{\partial x_i} + \sum_k \left(h_k \frac{\partial Y_k}{\partial x_i}\left(\frac{\lambda}{c_p} - \rho D_k\right)\right) \qquad \text{(Eqn. 6)}$$

The first term describes the conductive heat transport through Fourier's law (with $h = c_p T$, the heat capacity of the gas c_p and the thermal conductivity λ). The second term in eqn. 6 is the preferential species diffusion which describes the difference between mass diffusion and thermal diffusion. In a common approximation, the species diffusion coefficient is assumed to be equal for all species: $D_k = D$. This approximation is valid for most gas compositions of oxy-fuel or conventional combustion and does not only simplify eqn. 6 but also eqn. 4. However, the mixture-averaged diffusivity depends on the composition of the major species. An important approximation in combustion research assumes the Lewis-number $Le = \frac{\alpha}{D} = \frac{\lambda}{\rho D_i c_p}$ to be one. This is a valid approximation for a gas composition of similar molar mass M_i of all species. Then, the second term in eqn. 6 is equal to zero and the Schmidt- and Prandl-numbers Sc and Pr, which describe the ratio of momentum diffusivity to mass diffusivity and thermal diffusivity, are identical.[6] While in most cases the conductive and convective heat transfer within the gas phase has to be modeled and solved numerically, the heat transfer between the gas phase and solid surfaces in technical applications is often described through macroscopic semi-empirical correlations through the Nusselt-number $Nu = \alpha L/\lambda$, with the thermal diffusivity α and an application-dependent length-scale L. Using these correlations, the conductive heat transfer between the gas phase and e.g. the combustor walls or solid-fuel particles can be modeled.

For a holistic mathematical description of thermo-fluidic systems, constitutive equations of state have to be considered. These include e.g. the temperature dependent dynamic viscosity, heat conductivity and capacity and the thermal diffusivity. The relation between the temperature T, the pressure p and the density ρ in non-critical combustion systems can be described by the ideal gas law:

$$p = \rho \frac{R}{M} T \qquad \text{(Eqn. 7)}$$

with the universal gas constant R and the molar gas mass of the mixture $M = \sum \chi_i M_i = \left(\sum Y_i / M_i\right)^{-1}$.

---

[4] $Ma = \frac{|u_i|}{a}$, with the temperature-dependent speed of sound $a = \sqrt{\kappa R T / M}$ and the adiabatic exponent κ.

[5] While solid surfaces, such as combustor walls or coal particles, emit and absorb radiation by means of a continuous black-body radiator, which can be described by Planck's law, gases emit and absorb radiation through a complex multitude of discrete absorption lines, which have to be independently calculated (section 2.2).

[6] $Sc = \frac{\mu}{\rho D}$ and $Pr = \frac{\mu}{\rho \alpha}$

2.1.2 Turbulence

While the equations introduced in the last subsection allow to calculate the flow in most combustion applications in principle, they are practically impossible to be solved analytically in most applications due to their complex and coupled structure. Hence, expensive numerical simulations are required. One consequence of the complex nature of these differential equations is the necessity to distinguish between two regimes of flows: laminar and turbulent. In laminar flows minor perturbations are damped by the viscous forces and the flow field is kept regular. This characteristic allows one- or two-dimensional steady state flows. In turbulent flows, however, the perturbations are amplified, which leads to a multi-scalar, multi-dimensional and non-stationary flow field with a stochastic and quasi-random behavior in both space and time. Whether the flow is turbulent or laminar depends on its characteristic Reynolds-number:

$$Re = \frac{\rho U L}{\mu}$$
(Eqn. 8)

with characteristic velocity U and length scale L of the problem and the dynamic viscosity μ. Hence, the Reynolds-number is the ratio between inertial forces and viscous force. If this ratio exceeds a critical value, a transition from laminar to turbulent flow is initiated. Osborne Reynolds [58] noticed velocity gradients in flow-normal direction to be the origin of eddy-structures, which were initiated by the perturbations. In fully developed pipe flows, for example, stochastic eddies develop due to shear-stress in the near-wall region. For Reynolds-numbers below $Re \approx 2300$, these eddies are damped by viscous forces, for Re above this critical value, turbulence is initiated. Due to significantly enhanced mixing as well as transport of heat and mass, most technical combustion applications rely on turbulent flows.

For a simplified description of turbulence, a split into a mean and a variation term of the turbulent flow quantities has proven valuable:

$$\theta(t, x_i) = \bar{\theta}(x_i) + \theta'(t, x_i)$$
(Eqn. 9)

with the quantities' mean $\bar{\theta}(x_i) = \lim_{t \to \infty} \frac{1}{t} \int \theta(t, x_i) dt$. In L. F. Richardson's interpretation of turbulence [59], turbulent eddies are formed via energy transfer from the flow at large scales, break apart into smaller eddies due to inviscid processes and dissipate into heat at small length scales, where viscous damping is dominant due to a low Reynolds number (fig. 2.1). As the large-scale eddies are induced by the main flow, they are anisotropic. Throughout the eddy brake-up in the inertial subrange, information on the eddy orientation is lost, leading to isotropic eddies at the smallest scales.

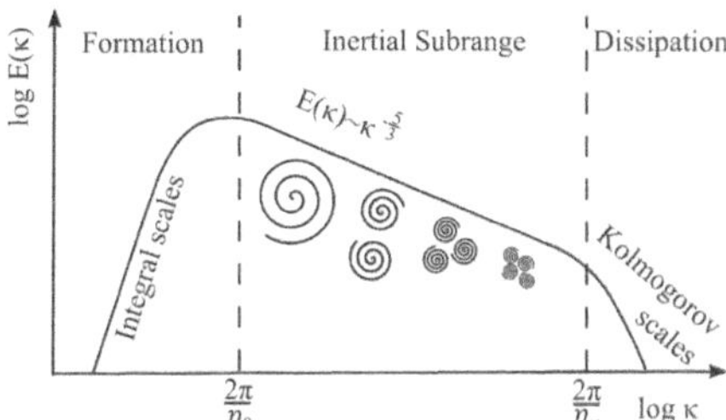

Fig. 2.1: Energy cascade of the turbulent energy spectrum

A. N. Kolmogorov [60] statistically found out that the energy spectrum $E_{TKE}(\kappa)$ of the turbulent kinetic energy

$$E_{TKE} = \frac{1}{2}\overline{u'_k u'_k} = \int E_{TKE}(\kappa)d\kappa \qquad \text{(Eqn. 10)}$$

in this subrange gets passed on to smaller scales through a power law dependency on the eddy wavenumber $\kappa = 2\pi/\eta$

$$E_{TKE}(\kappa) \sim \kappa^{-\frac{5}{3}} \qquad \text{(Eqn. 11)}$$

with the length scale of the eddy η. Hence, most of the kinetic energy is stored in the large scales. Two domains of scales can be identified: the integral scales of time τ_0 and length η_0 which separate the inertial subrange from the formation region, and the Kolgomorov microscales τ_K and η_k which differentiate between the inertial subrange and the dissipation region. While the first scales highly depend on the flow and can be calculated through two-point correlations, the latter only depend on the kinematic viscosity of the fluid ν_{fl} and the dissipation rate of turbulent kinetic energy $\varepsilon = 2\nu_{fl}\int \kappa^2 E_{TKE}(\kappa)d\kappa$:

$$\eta_k = \left(\frac{\nu_{fl}^3}{\varepsilon}\right)^{1/4} \qquad\qquad \tau_K = \left(\frac{\nu_{fl}^3}{\varepsilon}\right)^{1/2} \qquad \text{(Eqn. 12 and 13)}$$

The ratio between the integral and Kolmogorov scales can be estimated from the Reynolds-number.

2.1.3 Gas Combustion Principles

In combustion processes, the chemically bound potential energy of a fuel is converted into thermal energy through an exothermic oxidation reaction. In conventional combustion processes, the oxidizer, oxygen, is supplied through air, in oxy-fuel processes through a mixture with carbon dioxide. The chemical conversion of molecules is commonly depicted by macroscopic global reactions, which quantify the overall balance between reactants and products. However, a detailed view onto these global reactions reveals that the conversion consists of multiple underlying elementary reactions, which are necessary for a detailed calculation and description of the reaction rates, enthalpy release and formation of species. An example of the difference between the two descriptions is the combustion of methane in air: While the global reaction can be simply described by $CH_4 + 2\,O_2 \rightleftharpoons CO_2 + 2\,H_2O$, the underlying elementary sub-reactions are far more complex. The GRI-Mech 3.0 mechanism [61] for example which is already a reduced mechanism compared to the detailed chemistry, knows 325 elementary reactions and 52 species involved. The chemical forth and back reactions j with N species involved are generally depicted through:

$$\sum_{k=1}^{N} \varsigma'_{k,j}\, Sp_k \rightleftharpoons \sum_{k=1}^{N} \varsigma''_{k,j}\, Sp_k \qquad \text{(Eqn. 14)}$$

with the stoichiometric coefficients of the reactants $\varsigma'_{k,j}$ and products $\varsigma''_{k,j}$ and the chemical species Sp_k. These R chemical reactions allow to calculate the mass-fraction related species source term $\dot{\omega}_{i,Y}$ in eqn. 4 through the temporal change in the mole fraction χ_i of species i:

$$\dot{\omega}_{i,Y} = M_i\frac{d\chi_i}{dt} = M_i \sum_{j=1}^{R} (\varsigma''_{i,j} - \varsigma'_{i,j})\underbrace{\left(k_{f,j}\prod_{k=1}^{N}\chi_i^{\varsigma'_{k,j}} - k_{b,j}\prod_{k=1}^{N}\chi_i^{\varsigma''_{k,j}}\right)}_{\text{reaction rate } r_j} \qquad \text{(Eqn. 15)}$$

with the rate coefficients of the forth and back reaction $k_{f,j}$ and $k_{b,j}$, which can be derived from an Arrhenius approach:

$$k = AT^b \cdot \exp\left(-\frac{\Xi}{RT}\right) \qquad \text{(Eqn. 16)}$$

The rate coefficients depend on reaction-specific quantities A, b and the temperature T. A minimal activation enthalpy of Ξ is required to overcome the potential barrier of the reaction (fig. 2.2). At low temperatures, few reactants have a sufficient kinetic energy, and the likelihood of a reaction is low. Hence, even though the reactions are theoretically possible, the reaction rate is negligible compared to other reactions and practically no products are formed. Through the chemical conversion, an enthalpy of ΔH_j is released (or required for endothermic reactions), which allows to calculate the enthalpy source term $\dot{\omega}_{i,h}$ in eqn. 5 through $\dot{\omega}_h = \sum \Delta H_j r_j$ with the reaction rate r_j of the respective reaction being the last term in eqn. 15.

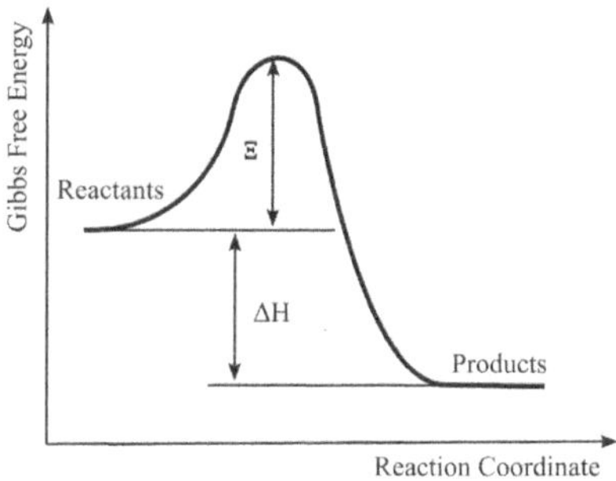

Fig. 2.2: Energy diagram for an elementary chemical reaction.

Commonly, flames that issue from chemical conversion are classified into different regimes (fig. 2.3). As the gas flow highly influences the flame structure and reaction rate, a first classification separates laminar from turbulent flame flows. As the reaction rates of laminar flames due to reduced mixing are typically significantly lower than those of turbulent flames, most technical combustion processes rely on turbulent flames to reduce the required combustion chamber dimensions. A chemical conversion and thus a flame requires the reactants fuel and oxidizer to be mixed and burned. Therefore, it is useful to distinguish flames based upon whether the fuel and oxidizer are mixed first and burned later, or if mixing and burning occur simultaneously. The first regime is referred to as premixed flames, the latter as non-premixed flames. For similar turbulence, the conversion rates of premixed flames typically exceed those of non-premixed flames, as the conversion rates are limited by the times scales of the quick chemical reactions rather than the time scales of mixing. However, some combustion processes require the reactants to be separated before combustion. Therefore, the reaction surface area between the zones of fuel and oxidizer are usually enlarged to achieve similar reaction rates, e.g. by creating small fuel droplets in a diesel engine.

Between these two regimes partially premixed or stratified flames exist, which typically rely on a fuel-rich primary flow and a lean, staged or purely-oxidizing secondary flow. These flames allow to reduce the formation of pollutants and to increase the flame stability, as they offer a broader influence on the flame thermochemistry and flow field. One example of a partially premixed flame is the RQL (rich burn, quick mix, lean burn) concept in the combustion chambers of modern jet engines [62]. The partially premixed flames show characteristics of non-premixed and premixed flows, depending on the application. For example, a rich Bunsen flame burns in a premixed mode, which is followed by a non-premixed diffusion-flame.

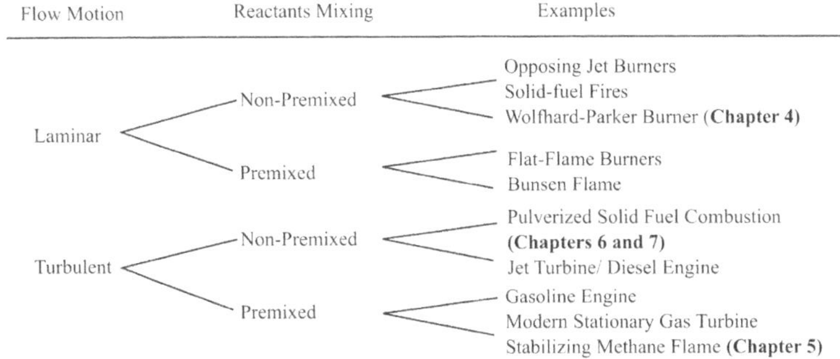

Fig. 2.3: Classification of flame types with examples including the flames investigated in this work with their respective chapters.[7]

In addition to the classifications through flow motion and reactants mixture, flames can be distinguished by their mixture composition through the equivalence ratio Φ, which is the ratio of the fuel/oxidizer mixture of the flame to stoichiometric conditions:

$$\Phi = \frac{Y_{Fuel}/Y_{Ox}}{(Y_{Fuel}/Y_{Ox})_{st}} = \frac{X_{Fuel}/X_{Ox}}{(X_{Fuel}/X_{Ox})_{st}} \qquad \text{(Eqn. 17)}$$

In stoichiometric conditions ($\Phi = 1$), all fuel and oxidizer are consumed. If more oxidizer is available than required by the fuel ($\Phi < 1$), the condition is referred to as fuel-lean, in the opposite case as fuel-rich ($\Phi > 1$). The equivalence ratio directly affects the temperature and species concentrations.

The mixture fraction ξ, which is a linear combination of the fuel and oxidizer mass fraction, can similarly be utilized to describe the chemical state of the mixture. For $\xi = 0$, the mixture is purely oxidizer, for $\xi = 1$, the whole mixture is composed of fuel. This quantity is advantageous, as, in contrast to the equivalence ratio, it yields finite values in pure fuel flows and is defined in the flue gas. From the mixture fraction, the equivalence ratio can be calculated through

$$\Phi = \frac{\xi}{1-\xi} \frac{1-\xi_{st}}{\xi_{st}} \qquad \text{(Eqn. 18)}$$

To calculate the mixture fraction, an elementary mass fraction $Z_j = \sum w_{ij} Y_j$ for each element needs to be defined, which is an analogy to the species mass fraction Y_i. w_{ij} denotes the mass proportions of the elements j in the species i. Then, ξ is defined through:

$$\xi = \frac{Z_i - Z_{i,Ox}}{Z_{i,Fuel} - Z_{i,Ox}} \qquad \text{(Eqn. 19)}$$

If the diffusivities of the elements are equal, the mixture fraction is independent from the choice of element i, as long as fuel and oxidizer do not have the same element mass fractions. Hence, a multitude of definitions of the mixture fraction exists. A simple definition through the element mass fractions of carbon Z_C and hydrogen Z_H in the fuel, $Z = Z_C + Z_H$, or the more complex Bilger mixture fraction[8] [63] are quite common. In the following, the characteristics of non-premixed and premixed flames will be discussed separately in brief.

---

[7] The stabilizing methane flame of chapter 6 is technically a partially premixed flame, as the primary flow is rich. However, the primary flow quickly mixes with a secondary flow, creating a lean mixture. Hence, in this chapter, the burner is treated as premixed for simplicity.

[8] $\xi_{Bilger} = \dfrac{\beta - \beta_{fuel}}{\beta_{oxidizer} - \beta_{fuel}}$ with $\beta = \dfrac{Z_C}{w_{C,fuel} \cdot M_C} + \dfrac{Z_H}{w_{H,fuel} \cdot M_H} - \dfrac{2 \cdot Z_O}{w_{O_2,oxidizer} \cdot M_{O_2}}$ and $w_{O_2,oxidizer} = w_{C,fuel} + (w_{H,fuel}/4)$

Non-Premixed Flames

In laminar or turbulent non-premixed flames, fuel and oxidizer propagate towards each other through convection and diffusion. Hence, the mixture fraction has a continuous transition from $\xi = 0$ on the oxidizer side to $\xi = 1$ in the fuel. Along the stoichiometric mixture, the flame is established. As the time scales of combustion are short compared to mixing, the chemistry is often described through the maxim "mixed=burned". This fact is depicted in the Burke-Schumann diagram (fig. 2.4), which shows the species mass fractions and temperature of an idealized non-premixed flame against the mixture fraction. At stoichiometric conditions ξ_{st}, the quantities exhibit extrema, and both fuel and oxidizer are completely consumed. From this reaction zone, the released heat and species diffuse into both directions. In the rich zone, due to the oxygen depletion, CO and unburned hydrocarbons from the flame are not consumed and their concentration is therefore high. In the fuel-lean zone, due to the oxygen-excess, the concentrations of the pollutant NO_x are high.

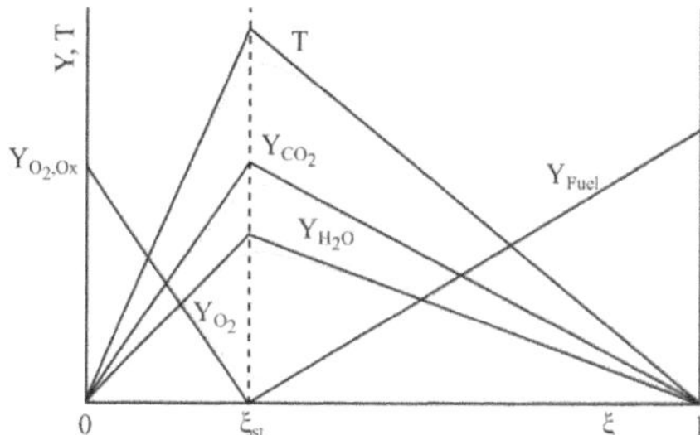

Fig. 2.4: Burke-Schumann diagram of species mass fractions and temperature against mixture fraction. [9]

Premixed Flames

In premixed flames, the reactants are perfectly mixed before approaching the thin reaction zone, where they are converted and consumed. Due to the perfect mixing, the fuel consumption of premixed flames is limited by the chemical reaction rates as opposed to the flow mixing in non-premixed flames.

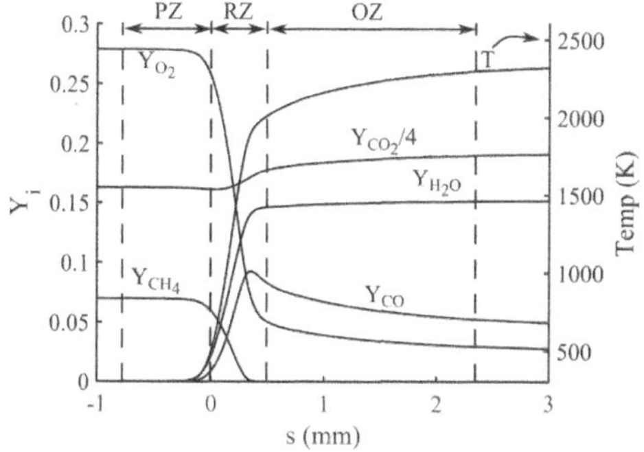

Fig. 2.5: Species mass fractions and temperature of a premixed, stoichiometric CH_4/oxy-fuel flame.[10]

---

[9] The linear relations between the mixture fraction and the species mass fractions only hold in an idealized system of mixed=burned and full reaction. In real systems, even at equilibrium conditions of a stoichiometric mixture, some reactants do not react due to the back-and forth equilibrium. In addition, finite flow strain rates pose deviations to the maxim mixed=burned. This effect is described with the help of the scalar dissipation rate, which is defined by $\chi \equiv 2D(\partial\xi/\partial x_i)^2$. Both effects are indicated by the grey lines in the figure.

[10] The simulations were performed by Sebastian Ganther (Energy and Power Plant Technology, TU Darmstadt) using Chem1D/GRI-Mech 3.0.

As the mixture fraction is a passive scalar calculated from conserved quantities, it is not altered during the combustion in premixed flames assuming equal diffusivities. Hence, the combustion is described with respect to a spatial coordinate s or a linearly increasing species mass fraction (e.g. a normalization of the H_2O mass fraction Y_{H_2O} with the equilibrium value) instead of the mixture fraction. Fig. 2.5 shows the species mass fractions of various molecules and the temperature over a spatial coordinate in a laminar, premixed and stoichiometric flame. The fuel of this flame, methane, is burned by the oxidizer O_2 which in this flame is supplied by an Oxy30 combustion atmosphere ($\chi_{O_2} = 0.3$; $\chi_{CO_2} = 0.7$).

The unreacted gas enters the negative side of the spatial coordinate. The laminar premixed flame can be separated into three regions: in the preheating or diffusion zone (PZ), intermediate species and combustion products from the reaction diffuse upstream. Additionally, the reactants are preheated through radiative or diffusive heat transport from the reaction zone and provide the activation energy for the primary chemical reactions. The high temperatures and the subsequently high reaction rates allow to stabilize a flame in the subsequent reaction zone (RZ), where most reactants are chemically converted into intermediates or products. As the reaction rates are high, the species mass fractions and the temperature exhibit large gradients in this zone. In the oxidation zone beyond the reaction zone, most of the reactions are complete. However, remaining intermediates, such as CO, are oxidized through slower reactions, which slightly increases the temperature further.

Two important quantities are necessary to classify premixed flames, the laminar flame speed s_L and the flame thickness δ_L. The laminar flame speed is the flow velocity of the reactants required to maintain a stationary premixed flame at one location, which is a reactant-specific quantity and depends on the equivalence ratio (fig. 2.6) and temperature.

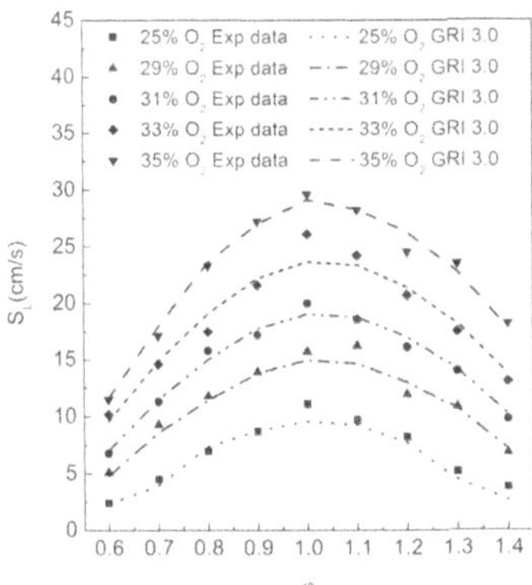

Fig. 2.6: Laminar flame speeds of methane in different oxy-fuel environments as a function of the equivalence ratio [11].

The laminar flame thickness δ_L is the thickness of the reaction zone, which is most commonly defined through the gradients of the temperature. As the reactions of premixed flames are limited by the quick chemical reactions only, the flame thickness at atmospheric pressures is usually small (~1 mm [14]).

In turbulent flames, the flame front is transported by eddies through turbulence-chemistry interaction, which causes wrinkling and stretching of the flame front. As this process allows to adjust the chemical conversion rate over a wide range through an increase of the flame surface by the turbulence, turbulent premixed flames are important for many technical applications, such as the gasoline engine. If the turbulence intensity is too high and the heat loss through heat transfer exceeds the generation through chemical conversion, the flame is locally quenched and torn.

The flame speed and thickness allow to classify the different flame regimes through the Borghi-Peters diagram (fig. 2.7). In this representation, the influence of the turbulence-chemistry interaction is evaluated through the ratio of the mean velocity fluctuation u' to the laminar flame speed s_L, u'/s_L, and the ratio of the integral length scale η_0 to the laminar flame thickness δ_L, η_0/δ_L. Several lines divide the diagram into domains with different flame behavior. When the turbulent Reynolds-number[11] Re_T is below one, viscous damping of the flow is dominant and it remains laminar. For a turbulent velocity fluctuations less than the laminar flame speed, the flame front is turbulently wrinkled. Above these lines, the flow is turbulent. However, when the flame thickness is less than the Kolmogorov scale, the system can be described as a locally laminar flame embedded in a turbulent flow, the so called flamelets (island formation). For a flame thickness above the Kolmogorov length scale, eddies are able to penetrate the flame and locally tear it. These two domains are separated by a unity ratio of these length scales, the turbulent Karlovitz-number $Ka = \frac{\eta_K}{\delta_L} = 1$. An additional distinction is observed when the time needed for chemical reaction exceeds the time needed for fluid motion, which means that most eddies are located inside the reaction zone. This corresponds to a Damköhler-number of unity $Da = \frac{\tau_0}{\tau_L} = \frac{\eta_0/u'}{\delta_L/s_L} = 1$ which seperates the torn flames from a well-stirred reactor domain.

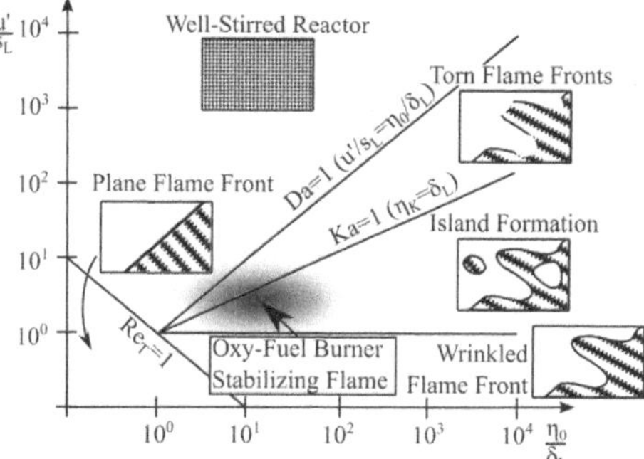

Fig. 2.7: Borghi-Peters diagram of premixed combustion regimes with the approximate classification of the stabilizing methane flames of chapter 5.

As the WHP-burner and the solid-fuel flames are globally or locally non-premixed flames, the stabilizing methane flames (chapter 5) are the only (partially-) premixed flames[7] which can be located in the diagram (fig. 2.7)[12]. The location indicates that most of the flame front is closed but can break into islands. This is supported by observing the OH-images of [23].

More information on gas combustion principles can be found in [65–67]. The figures of this subsection are inspired by [65], the extension of the balance equations by the phase exchange terms P by [68].

2.1.4 Solid Fuel Combustion Principles

Peat, charcoal, lignite, biutuminous coal, anthracite and biomass, such as wood, straw and crop, are examples of solid fuels of primarily organic constituents (C, H, N, O and S-compounds). The organic material, which contains the energy content of the fuel, is mixed with inorganic minerals, such as chlorine compounds, silicon oxides and metal oxides, which mainly originate from either plant life (biomass and peat) or from extraneous mineral matter (lignite and coal) such as clay. Lignite and coal developed from biomass in a process of partial decomposition under air-deficient conditions. The biomass accumulated on land and in swamps during previous geological periods and became compacted and dewatered by the pressure of overlaying sediments. At greater depth, higher temperature, pressure and age, lignite develops from peat, and turns into bituminous coal and eventually anthracite. During

---

[11] $Re_T = \rho u' L / \mu$

[12] To determine the location of the flames in the diagram, u' and η_0 were estimated from [23], s_L from [11] and [64] and δ_L from [14].

this coalification process, the moisture is reduced while the solid carbon content and thus the heating value and rank are increased. Additionally, lightly bound volatiles, which are a mixture of mainly short- and long-chain hydrocarbons, aromatic hydrocarbons and some sulfur, are either oxidized or gassing out, which reduces the oxygen and hydrogen contents. This process is illustrated in the van Krevelen diagram (fig. 2.8), which is used to classify the origin and maturity of solid fuels based on the atomic ratios of hydrogen and oxygen to carbon. The diagram allows to distinguish between typical domains of the different solid fuels, whose rank, age and density are increased from the initial biomass, which mainly consists of cellulose, hemicellulose and lignin [69,70], to anthracite, which contains large shares of graphite.

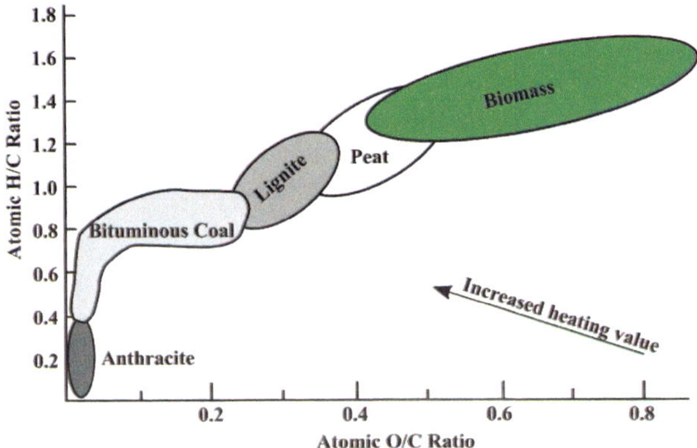

Fig. 2.8: van Krevelen diagram of solid fuels.

As the chemical structure of solid fuels is complex and consists of large macromolecules, the chemical composition is commonly assessed by means of proximate- and ultimate analyses measuring the mass fractions of the chemical elements, such as C, O, N, H, S and Cl, in combination with the moisture, ash and volatile content as well as the heating value. The microstructure of solid fuel molecules consists of polymer networks of short and long hydrocarbons with peripheral functional groups in combination with aromatic clusters, which form side chains and loops and are interconnected by bridges. With increasing rank and coalification, the molecules become larger and denser and aromatic clusters as well as graphite more dominant. The content of functional groups which contain a large fraction of the O- and N-compounds reduces due to devolatilization.

Prior to combustion in conventional power plants, solid fuels are pulverized in mills to particle diameters of 10–100 µm. The heterogeneous combustion process of these particles is more complex than the homogeneous gas combustion, which is described in the previous subsection. Fig. 2.9 schematically depicts the combustion process of solid fuels in pulverized fuel firing. At first, the high gas temperature in the ambience of the particles vaporizes its H_2O content, which causes a drying of the particle. With increasing temperature, other gases which have formed during the coalification process, such as carbon dioxide and nitrogen, outgas as well. The fraction of these gases that desorb into the porous structure of the particle expands the softened particle at higher temperature thermally and initiates swelling. In addition to swelling, the increasing temperatures lead to a formation of tars and burnable gaseous products, such as CO, CH_4, C_2H_2, C_2H_4 and C_2H_6, due to cracking of the chemical compounds of the fuel. This process is referred to as devolatilization or, in the absence of O_2, pyrolysis. The volatiles leave the particle through the pores and form a volatile atmosphere around the particle. After ignition, the volatiles undergo a chemical conversion which can be described by the homogeneous reactions of a non-premixed flame. The diameter of the flame enveloping a particle is about three to five times the diameter of the particle [71]. In fig. 2.9, this flame is simplified by a laminar sphere, which motivates the investigation of the corresponding flame depicted in the generic WHP-burner experiment in this work. For higher turbulence, the gases mix and eventually undergo a transition into the partially premixed regime, which is similar to the stabilizing methane flame investigated in chapter 5 of this work.

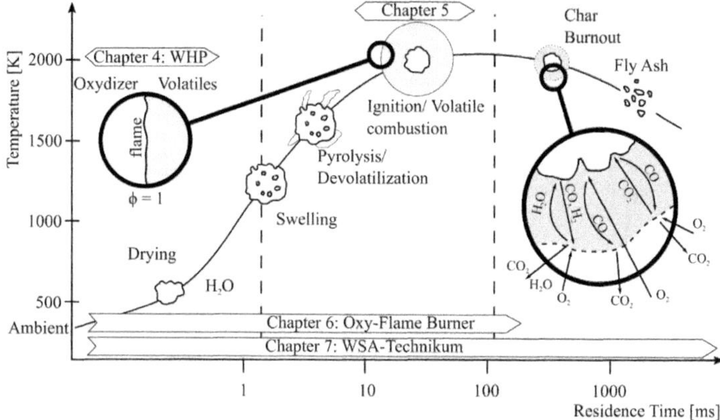

Fig. 2.9: Schematic of the combustion process in pulverized solid fuel firing as a function of the particle residence time and temperature.

The heat necessary for drying, pyrolysis and ignition is transferred mostly by gas convection or radiation from particle or wall surfaces as well as the triatomic gas molecules. After devolatilization of the particle, a porous char structure of mainly carbon and ash remains, which is oxidized at sufficiently high temperatures. This heterogeneous process of char combustion is by several orders of magnitude slower than the homogeneous volatile combustion. The insert in fig. 2.9 depicts three different global mechanisms that allow a chemical conversion of the solid fuel at the char surface:

- $C + 0.5\ O_2 \rightleftharpoons CO \quad + 111\ kJ/mol \quad$ (heterogeneous oxidation)
- $C + CO_2 \rightleftharpoons 2\ CO \quad + 172\ kJ/mol \quad$ (Boudouard reaction)
- $C + H_2O \rightleftharpoons CO + H_2 + 131\ kJ/mol$

Each of the exothermic reactions forms CO, which in combination with CO_2, H_2O and O_2 generates a gaseous atmosphere around the particle. The oxidants have to penetrate this atmosphere boundary layer through diffusive transport to allow a homogeneous reaction of the products CO and H_2 through the reactions $CO + 0.5\ O_2 \rightleftharpoons CO_2$ and $H_2 + 0.5\ O_2 \rightleftharpoons H_2O$. Hence, the rate coefficients of the conversion are limited by the highest of the times required for chemical reaction or for transport, which in turn can be separated in pore diffusion or boundary layer diffusion. The three domains are depicted in an Arrhenius diagram in fig. 2.10. At low temperatures (I), the reaction on the surface of the particle is slow and thus limits the overall reaction.

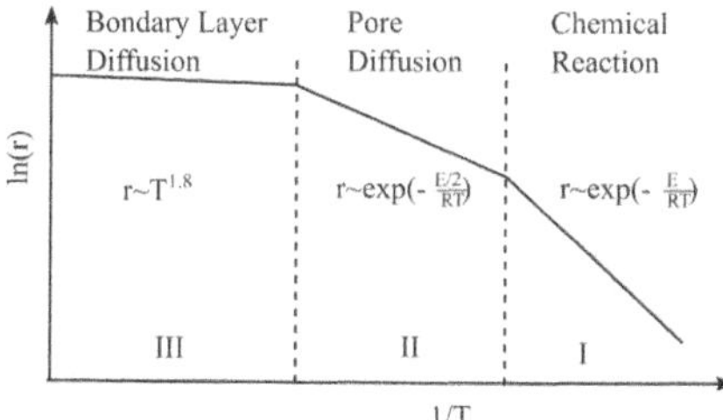

Fig. 2.10: Rate coefficient of char combustion as function of the inverse temperature [72].

At higher temperatures (II), these reactions become more intense such that the diffusion of oxidizer into the pores becomes the limiting factor of the overall reaction. However, the atmosphere around the particle still contains high amounts of oxidizer. At even higher temperatures, these molecules are quickly consumed as well, which limits the overall reaction rate to the diffusion rate required to penetrate the boundary layer (III).

The solid fuel flames at the Oxy-Flame Burner and the Technikum allow a combustion of coal particles through the entire consumption in fig. 2.9. Yet, the extent of char combustion in the Oxy-Flame Burner is unclear due to the short residence time in the hot reaction zone. These flames are investigated in chapters 6 and 7.

Insightful further information on the combustion of solid fuels can be found in [72–74].

2.1.5 Oxy-Fuel Combustion

A transition from an air-blown combustion of fossil fuels to a conversion under oxy-fuel combustion atmosphere is accompanied by major changes in the combustion behavior. The oxy-fuel combustion atmosphere has an increased molar heat capacity, which leads to a lower adiabatic flame temperature and a decreased laminar flame speed for a given oxidizer concentration in the combustion atmosphere. These alterations eventually influence the flame stability. The change of temperature has a significant impact on the plant thermodynamics and the reaction rates. Additionally, the mass diffusion coefficient D_{O_2} of oxygen is $\approx 30\%$ lower in a CO_2 gas matrix than in an N_2 atmosphere. This causes an increased Lewis number and a need for higher oxygen concentrations [32]. The increased gas density influences the momentum flux and thus the flow field, which, besides the implications on the combustor, influences the compressor and turbine design in oxy-gas combustion. As the kinematic viscosity μ of CO_2 is approximately half of the N_2 value, the flow Reynolds number increases and leads to higher turbulence. Due to the high concentrations of tri-atomic molecules, the radiative heat flux is increased [75]. This affects the heating rate of the solid fuel particles, the heat release and eventually the design of the combustor, superheater and heat exchanger. Furthermore, in contrast to nitrogen, carbon dioxide is not inert during the combustion as it reacts to CO at higher temperatures via the water-gas shift reaction ($CO_2 + H_2 \rightleftharpoons CO + H_2O$) and Boudouard reaction, which was introduced in the previous subsection. The higher CO concentrations, in return, influence the char burnout, which locally generates a CO-enriched atmosphere around the char particle in oxy-coal combustion. A flue gas recirculation, which is required for a closed-loop plant operation, further increases the complexity due to the increased water vapor content in the recycled gas and the concentrations of CO, NO_x, sulfur oxides, HCl and particles.

Besides the combustor, one of the greatest challenges and delimiting factor to the plant efficiency of oxy-fuel combustion systems is the air separation unit. Current techniques rely on oxygen production through cryogenic distillation, diffusion through polymeric membranes or pressure swing adsorption. These units are both expensive and energy-intensive, particularly for achieving high O_2-purities. Beyond 97% purity, the energy required for further purification increases exponentially [76]. Hence, practical oxy-fuel power plants are expected to be operated using impure oxygen. In addition, the CO_2-content in the combustion atmosphere is contaminated by residuals of the combustion, such as the aforementioned pollutants and O_2, as well as N_2 from air leakages. Through looping of flue gas, the concentrations of these species accumulate, which reduces the purity of CO_2 in the exhaust.

Independent of the type of power plant (i.e. gas turbine or condensation steam power plant firing solid fuels), the oxygen content of the air needs to be separated prior to mixing with recirculated flue gas. Afterwards, the mixed gases react with the fuel in the combustors of either solid fuel steam power plants (I in fig. 2.11) or gas turbines (II). In gas turbines, the gas is initially pressurized through the compressor. A pressurized oxy-fuel combustion is also possible for solid-fuel steam power plants and can have a positive impact on the overall plant efficiency, as the energy required for pressure-liquefaction is lowered. After the combustion, a fraction of the flue gas is split apart for flue gas recirculation before the water vapor content is removed in the condenser.

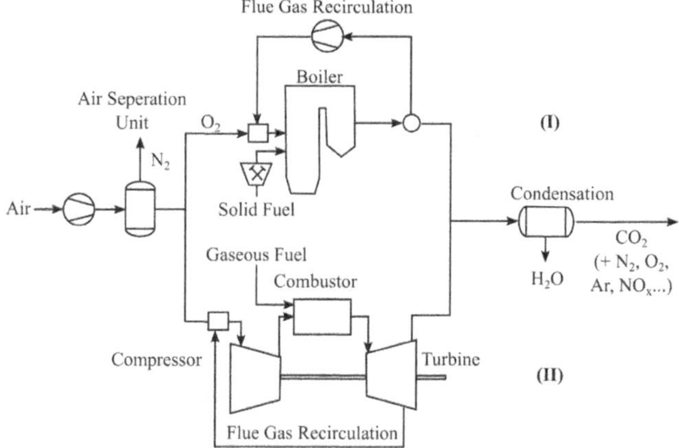

Fig. 2.11: Process diagrams of (I) an oxy-coal steam power plant and (II) an oxy-fuel gas turbine (inspired by [73]).

Both the solid-fuel steam power plants and the gas turbines can eventually be retrofitted for oxy-fuel combustion.

2.2 Molecular Spectroscopy

For many years, optical measurement techniques, in particular laser-based methods, have developed into state of the art in combustion research. The techniques are advantageous as they

- provide a higher selectivity, sensitivity, temporal and often spatial resolution than other methods,

- are minimally invasive and allow an in-situ analysis of multiple parameters,

- can be set up to measure a wide range of quantities, in many cases yielding quantitative results.

All of these techniques are based on the interaction between light and matter. While the classical concept of light as an electromagnetic wave allows to describe optical phenomena such as dispersion, scattering, reflection, refraction and diffraction through the Maxwell-equations, it fails to explain effects such as fluorescence or the Fraunhofer lines in the solar spectrum [77]. These effects can only be described quantum-mechanically through a quantization of states in matter and light, which will be explained in this section together with the spectroscopic principles of the measurement technique utilized in this work.

In 1900 Max Planck postulated an energy-discretization of light [78]. In addition to the conception of light as a wave, this theory describes light as particles with zero rest mass, the photons, whose energy is linearly dependent on the lights frequency f:

$$E = h{\cdot}f = h{\cdot}c{\cdot}v \qquad \text{(Eqn. 20)}^{13}$$

with the Plank constant h (h=6.626·10^{-34} J·s). In contrast to the description as a wave, the energy of light seen as a particle can only take multiples of this minimal energy. According to this finding, light has characteristics of both a wave and particle, which is known as wave-particle duality. In return, Werner Heisenberg, Louis-Victor de Broglie, Albert Einstein and Niels Bohr abandoned the classical atomic models and attributed particles with characteristics of waves [79–82]. Similar to an energy-discretization of photons, Bohr postulated energetic states in atoms to occur only at discrete energy levels. Accordingly, energy transfer from an initial state $\langle i|$ to a final state $\langle k|$ is discontinuous and only possible for discrete energy differences $\Delta E = E_k - E_i$. This model explained the occurrence of discrete lines in the spectrum of light emitted by gases, in particular by hydrogen.

Erwin Schrödinger found a way to calculate the energy of states in quantum-mechanical systems, which include both photons and matter, by solving the partial differential equation:

$$\hat{H}\psi = E\psi \qquad \text{(Eqn. 21)}$$

where $\hat{H}$ is the Hamilton operator of the system which combines the operators for the system's potential V (e.g. Morse-potential) and the kinetic energy of its components, i.e. the s electrons and k atomic nuclei:

$$\hat{H} = -\frac{\hbar^2}{2}\left(\underbrace{\frac{1}{m_e}\sum_{s}^{t}\Delta_s}_{\substack{\text{electron}\\\text{motion}}} + \underbrace{\sum_{j}^{k}\frac{1}{M_j}\Delta_j}_{\substack{\text{nuclei}\\\text{motion}}}\right) + V \qquad \text{(Eqn. 22)}^{14}$$

Here, m_e denotes the mass of the t electrons, M_j the masses of the k nuclei and Δ the Laplace-operator. The solution of the Schrödinger equation, the imaginary wave function ψ, is not a physical observable, but its square $|\psi|^2$ is real and describes the probability density of the system for a given spatial coordinate and energy level. The eigenvalues E of the solution are the system's energy levels. For simple systems,

---

[13] In spectroscopy, the wavelength λ=c/f (in nm) and wavenumber v=λ^{-1} (in cm^{-1}) are more commonly used than the frequency notation.

[14] $\hbar = h/2\pi$

such as the hydrogen-atom with only one valence electron, the Schrödinger equation can be solved analytically. For multi-body systems, such as molecules, approximations and often simulations are necessary.

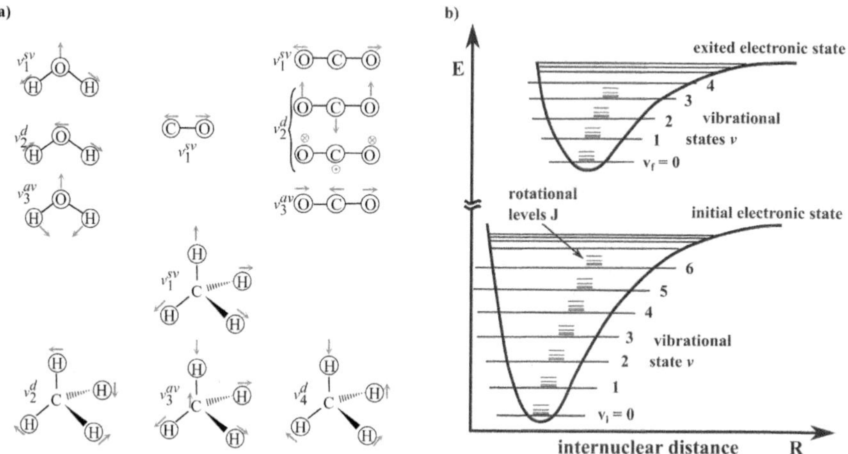

Fig. 2.12: a) Vibrational normal modes of CO, H_2O, CO_2 and CH_4 [15]; b) Rotational and vibrational levels in two different electronic states of a molecule (similar to [83]).

In the Born-Oppenheimer approximation, the motions of nuclei and electrons of gas molecules in a covalent bound are separated from each other. As their masses are largely different, the motions of electrons are simplified to occur instantaneously compared to the motions of the nuclei. In principle, the motion in molecules can be subdivided into three primary degrees of freedom:

- The rotational states originate from a rotation of the massive nuclei around the molecular axis and center of mass. Non-linear molecules have three rotational degrees of freedom, linear molecules two.

- A relative motion of nuclei to each other around the internuclear equilibrium distance results in a vibrational state. Non-linear molecules of N nuclei have 3N-6 vibrational degrees of freedom, linear molecules 3N-5. The normal modes of exemplary molecules are shown in fig. 2.12a.

- The electronic states originate from the motion of electrons within orbitals around then nuclei, and are caused by the electrons' spin and the Coulomb potentials of the nuclei and the additional electrons.

Each state is described by a set of quantum numbers. For example, the energetic levels of a diatomic molecule´s rotational states can be calculated to:

$$E_{rot} = \frac{J(J+1)\hbar^2}{2MR^2}$$

(Eqn. 23) [16]

---

[15] The upper indices denote the symmetrical stretching vibration (sv), asymmetrical stretching vibration (av) and the deformation vibration (d).

[16] For a rigid rotator, the rotational axis R corresponds to the equilibrium distance R_0 between the molecules. Considering a centrifugal extension of the rotational axis $R = R_0 \left(1 + \frac{J(J+1)\hbar^2}{M k R_0^4}\right)$ which is increasing with the angular momenta J, a Taylor-expansion allows to calculate the rotational energetic levels.

with the reduced mass of the molecule M and the rotational quantum number $J=0, 1, 2,...$. The vibrational energy levels of an oscillator in an inharmonic Morse-potential calculate to:

$$E_{vib} = \hbar\omega_0 \left(v + \frac{1}{2}\right) - \frac{\hbar^2\omega_0{}^2}{4E_D}\left(v + \frac{1}{2}\right)^2 \qquad \text{(Eqn. 24)}$$

with the Morse-potential's dissociation energy E_D and the ground-state annular frequency ω_0 as well as the vibrational quantum number v. [17]

Due to the large number of degrees of freedom, potentials and necessary correction terms for non-simplified theories and couplings, molecules of higher complexity require a large set of quantum numbers to describe the molecular state. These calculations exceed the scope of this work. The details on the calculation of the quantum numbers and their selection rules can be found in [83–85]. As this work primarily aims for rotational-vibrational spectroscopy, the most important quantum numbers are the rotation's total angular momentum J and the vibrational quantum number v.

A change of state requires an energy difference $\Delta E = E_k - E_i$, which depends on the quantum numbers of the initial and final states. Similarly to the motion, the total energy of the molecule can be separated into the rotational energy E_r, the vibrational energy E_v and the electronic energy E_e, which generally follow the relation $E_r << E_v << E_e$. For this reason, each electronic state can be subdivided into multiple vibrational states which in turn can be subdivided into rotational states (fig. 2.12b). Transitions between rotational states for constant vibrational and electronic states typically have wavenumbers in the magnitude of $v{\sim}10$ cm^{-1}. The entirety of possible purely rotational transitions causes bands of lines in the microwave spectrum. Transitions between vibrational states without alteration of the electronic state provoke bands in the infrared spectrum with typical wavenumbers of $v{\sim}1000$ cm^{-1} (e.g. H_2: up to 4160 cm^{-1}, N_2: up to 2300 cm^{-1}). Electronic transitions range from the visible to the ultraviolet spectrum, with wavenumbers in the order of $v{\sim}20\ 000$ cm^{-1}.

In resonant interactions between light and matter, the energy difference between the states is equal to the energy of the photon involved: $\Delta E = E_k - E_i = h \cdot c \cdot v$. According to Einstein, energy transfers can occur in three ways (fig. 2.13):

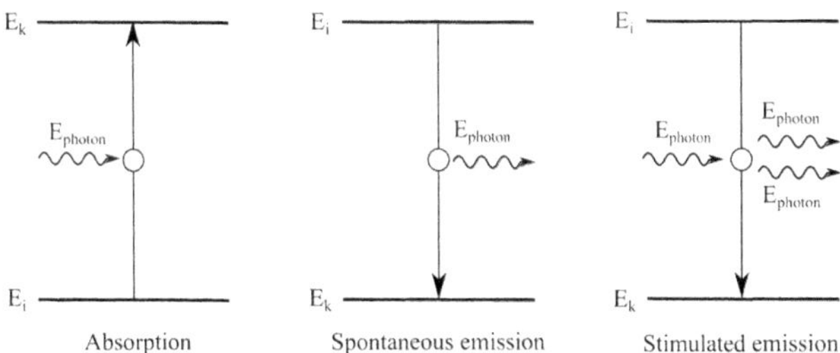

Fig. 2.13: Schematic representation of resonant light-matter interaction between a photon and a particle (atom or molecule) in a simplified two-level system.

- **Absorption:** Initially, the particle is in the lower energetic state. If the photon energy corresponds to the energetic difference to the upper state, the photon can transfer the particle into the upper state. As the photon is annihilated during this process, the intensity of the light is reduced at the wavelength corresponding to the photon energy, which results in absorption line spectra.

- **Spontaneous emission:** If the particle is originally in the upper energetic state, it can undergo a random transition to the lower state energy. The energetic difference is emitted by a photon,

---

[17] $\omega_0 = a\sqrt{2E_D/M}$. a is a constant of the Morse-potential

which leads to emission spectra. The average time a particle remains in the upper energetic state, the lifetime, is usually a matter of nanoseconds but can reach milliseconds and even seconds for forbidden transitions (i.e. particle fluorescence).

- **Stimulated emission:** If a particle in the upper energetic state interacts with a photon before undergoing spontaneous emission, an induced transfer into the lower state can be provoked. Hereby, a second photon of identical wavelength, polarization and phase is emitted. This is the fundamental principle of laser light sources (light amplification by stimulated emission of radiation).

In addition to the three transitions, non-radiating transitions are possible, e.g. through collisional energy transfer with other particles or emission of valence electrons (Auger-effect).

2.2.1 Absorption Line Strength and Position

While the eigenvalues of the Schrödinger equation (eqn. 21), the energetic levels of each state, allow to calculate the absorption line positions via $v_{0,i,j} = (E_k - E_i)/hc$, they do not yield information on the population of a state or the transition probability. The probability per second for a single excited molecule to undergo a spontaneous emission from an excited state $\langle i|$ to an energetically lower state $k|$ is given by the Einstein-coefficient for spontaneous emission A_{ik}. Likewise, the wave function needs to undergo a transition $\psi_i \rightarrow \psi_k$ for a spontaneous emission. Hence, the Einstein-coefficient depends on the secondary diagonal element of the transition moment M:

$$M_{ik} = e \int \psi_i^* \mathbf{r} \, \psi_k \mathrm{d} \mathbf{r} \qquad \text{(Eqn. 25)}$$

As the indices i and k are abbreviations for the quantum numbers of each state, the integration spans the entire configuration space of states of the electrons and nuclei, whereby the Born-Oppenheimer again allows a separation into electronic and rotational-vibrational terms. Due to the integration over the electronic and nuclei coordinates, a transition from one energy level to another is more likely to happen if the wave functions overlap more similarly, which is known as the Franck-Condon principle (fig. 2.14) of vibrational transitions.[18] The Einstein-coefficient for spontaneous emission and the transition moment are related via:

$$A_{ik} = \frac{1}{g_2} \frac{2(2\pi v)^3}{3\varepsilon_0 h} |M_{ik}|^2 \qquad \text{(Eqn. 26)}$$

with the vacuum permittivity ε_0 and the wavenumber of the emitted photon v. g_2 is the degeneration of the excited state and counts the states with identical energy levels but different sets of quantum numbers. For example, the deformation vibrations of CO_2 in fig. 2.12a are double-degenerated, as the orientation of the molecule with respect to the deformation axis does not change the energy. Likewise, g_1 is the degeneration of the lower state. The transition rate of a single molecule to undergo a transition from an energetically lower to an excited state is given by the Einstein-coefficient for absorption, B_{ki}, which is related to the transition moment through:

$$B_{ki} = \frac{g_2}{g_1} \frac{1}{(8\pi h v^3)} |M_{ik}|^2 \qquad \text{(Eqn. 27)}$$

The relation to the Einstein-coefficient for induced emission, B_{ki}, and the oscillator strength can be found in [86,87]. As individual transitions are visible as sharp lines in the spectra, they are often referred to as absorption or emission lines. For an ensemble of gas molecules, the line strength S of each transition depends, besides the transition rate, on the fractional population of the states. In thermal

---

[18] For rotational transitions, the overlap of the wave functions is known as the Hönl-London factor.

equilibrium, the fraction of molecules in a state $\langle i|$, N_i, to the overall number of molecules $N = \sum_i N_i$ is given by the Boltzmann distribution:

$$\frac{N_i}{N} = \frac{g_i}{Q(T)} \cdot e^{-\frac{E_i}{k_B T}} \qquad \text{(Eqn. 28)}$$

with the degeneration and energy of the state g_i and E_i, the Boltzmann-constant k_B and the gas temperature T. The partition sum $Q(T)$ is a normalization over all states and given by:

$$Q(T) = \sum_i g_i e^{-\frac{E_i}{k_B T}} \qquad \text{(Eqn. 29)}$$

In the Born-Oppenheimer approximation the partition sum can again be separated into contributions from rotational, vibrational and electronic states. At ambient temperatures, the thermal energy $k_B T \approx 200 \text{ cm}^{-1} \cdot hc$ exceeds the energies of typical rotational transitions but undercuts those of electronic and vibrational transitions. Thus, the population of the rotational states is Boltzmann-distributed, whereas by far most molecules are in the electronical and vibrational ground states. The transitions from the vibrational ground state by $\Delta v = +1$ result in the strong vibrational fundamental bands, whereas transitions to and from higher vibrational states are referred to as overtones. In general, all rotational transitions within a vibrational state by $\Delta J = +1$ are known as the lines of the R-branch, those by $\Delta J = -1$ as P-branch. At temperatures typical for combustion environments, the thermal energy is in the order of magnitude of the vibrational transition energy and higher vibrational states are populated and lead to additional strong hot-bands.

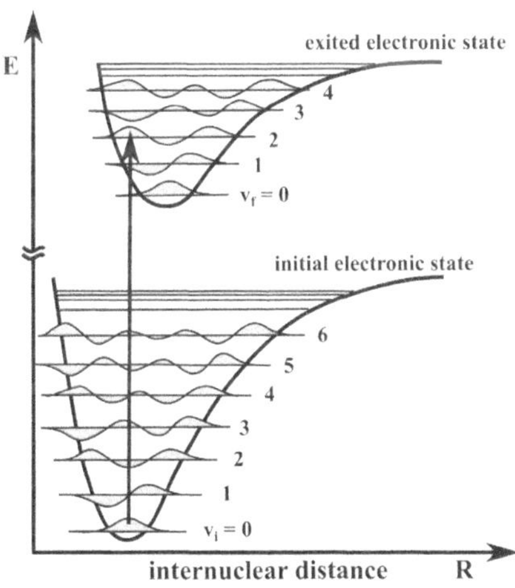

Fig. 2.14: Wave-functions of the first vibrational states within an electronic state which can be described as Hermite polynomials; Franck-Condon-principle.

From the transition probability and population fraction, the line strength $S(T)$ for absorption can be calculated by [87]

$$S(T) = \frac{N_i}{N} \frac{B_{ki}}{h\nu} \qquad \text{(Eqn. 30)}$$

For a simplified handling in databases, the absorption line strength is often calculated for a reference temperature T_{ref} (in the most common databases HITRAN and HITEMP [88,89], a reference temperature of $T_{ref} = 296$ K is used). Then, the line strength $S(T)$ can be calculated from the reference value $S(T_{ref})$, the energy of the lower state E_i and the partition sum $Q(T)$:

$$S(T) = S\big(T_{ref}\big)\frac{Q(T_{ref})}{Q(T)} \cdot e^{-\frac{E_i}{k_B}\left(\frac{1}{T}-\frac{1}{T_{ref}}\right)} \cdot \frac{1 - e^{-\frac{hcv_0}{k_BT}}}{1 - e^{-\frac{hcv_0}{k_BT_{ref}}}} \qquad \text{(Eqn. 31)}$$

In databases, the partition sum is in turn filed for a reference temperature $Q(T_{ref})$ to reduce the complexity. The temperature dependency can be calculated from empirical polynomials. More information on the determination on the reference line strength used in this work can be found in section 3.5.

2.2.2 Line Shaping and Shifting Effects

Even though the energy levels of states in a molecule are discrete, the absorption lines caused by transitions between the states are non-singular, as, due to several physical effects, the lines are broadened and shifted. The most relevant effects for combustion environments will be explained in the following. Further details can be found in [84–87].

Natural Line Width

In a semi-classical model, dipole oscillations in a harmonic potential lead to an exponential reduction of the amplitude due to the radiative energy loss. A Fourier-transformation from the temporal to the wavenumber space lead to a Lorentz-distribution of the intensity

$$\phi_n(v - v_{0,i,j}) = \frac{\Delta v_{n,i,j}/2\pi}{(v - v_{0,i,j})^2 + \big(\Delta v_{n,i,j}/2\big)^2} \qquad \text{(Eqn. 32)}$$

Thereby, ϕ_n describes the normalized natural line shape around the absorption line central frequency $v_{0,i,j}$ of transition i and species j, and $\Delta v_{n,i,j}$ the corresponding natural line width[19]. In a quantum-mechanical description, the lifetime of the excited state $\tau_{i,j}$ together with its complementary variable $\Delta E = E_i - E_k$ are subject to Heisenberg's uncertainty principle [82]

$$\Delta E \tau_{i,j} = hc \cdot \Delta v_{n,i,j} \cdot \tau_{i,j} \geq \frac{\hbar}{2} \qquad \text{(Eqn. 33)}$$

Thus, the line width is non-singular and related to the lifetime of the excited state via $\Delta v_{n,i,j} = 1/\big(2\pi c \cdot \tau_{i,j}\big)$. The natural line is too small to play a relevant role in technical applications, as it is covered by several orders of magnitude stronger broadening effects.

Doppler Broadening

In an ensemble of molecules, the thermal motion[20] of molecules induces a Doppler shift of the emission or absorption frequency. If the particle is moving towards the observer, the absorption wavelength is blue-shifted, in the opposite direction, a red-shift occurs. Since the particle velocities, due to the thermal distribution of the kinetic energy, as well as the particle motion directions show a distribution, a continuous broadening of the absorption line is induced. The broadening distribution is equivalent to a Gauss-profile of:

---

[19] The line width is defined as the full width at half maximum, FWHM
[20] The mean thermal velocity of molecules is $\bar{v}_{th} = (2k_BT/\pi M_j)^{0.5}$. For example, the mean thermal velocity of H_2O at 1500 K is $\bar{v}_{th} \cong 700$ m/s.

$$\phi_D(v - v_{0,i,j}) = \frac{1}{\Delta v_{D,i,j}} \sqrt{\frac{4\ln(2)}{\pi}} \, exp\left(-4\ln(2)\left(\frac{v - v_{0,i,j}}{\Delta v_{D,i,j}}\right)^2\right) \qquad \text{(Eqn. 34)}$$

Thereby, the factor $\Delta v_{D,i,j} = \frac{v_{0,i,j}}{c}\sqrt{8k_B T \ln 2 / M_j}$ is the Doppler-broadening of transition i and species j, which only depends on the gas temperature and the molecular mass M_j. Hence, the Doppler broadening can be calculated with high accuracy. For rotational-vibrational transitions at ambient and combustion-relevant temperatures, the Doppler-broadening is in the order of $10^{-2} - 10^{-1}$ cm^{-1}.

Collisional Broadening and Shift

Two approaching molecules mutually alter their potential due to the energy-dependent particle-particle interaction, which induces a shift in the energy levels. This energy shift depends on the orbital structure of both molecules and their inter-nuclear distance and can be positive for repulsive potentials between the constituents and negative for an attractive potential. While elastic collisions only induce a line broadening and shift, an inelastic collision additionally alters the population density of each state, which changes the line strength. These quenching collisions are negligible for infrared absorption lines due to the short lifetime of the excited state, but can have tremendous influence on long-living fluorescence lines. In a semi-classical model, the collisions induce a phase shift in the absorbed or emitted electromagnetic wave. Depending on the model of the collision (soft or hard model), the broadening leads to different line shapes [90]. Mostly, the collisional broadening is modeled by means of a Lorentzian distribution (eqn. 32). The impact of molecules of the same species as the target molecule is referred to as self-broadening and is separated from the impact of foreign gases, the foreign-broadening. The total collisional broadening coefficient $\Delta v_{C,i,j}$ of the distribution can be calculated to:

$$\Delta v_{C,i,j} = \gamma_{i,j,s} p_s \left(\frac{T_{ref}}{T}\right)^{n_{i,j,s}} + \sum_f \gamma_{i,j,f} p_f \left(\frac{T_{ref}}{T}\right)^{n_{i,j,f}} \qquad \text{(Eqn. 35)}$$

The first term in this equation describes the self-broadening by the same species as the emitting or absorbing molecule s. In the second term, the foreign-broadening is depicted, which is summed over all foreign species f. Each term itself consists of two factors: The first part is the pressure dependency which depends on the partial pressures p_s or p_f and broadening coefficients γ_s or γ_f. Generally, these coefficients are transition-dependent. The second part is the temperature dependency which requires the reference temperature T_{ref} and an exponent $n_{i,j,s}$ or $n_{i,j,f}$, which is the temperature-coefficient. As the self- and foreign broadening coefficients are usually in the magnitude of $10^{-2} - 10^{-1}$ cm^{-1}/atm, the collisional and Doppler width are in the same order of magnitude.

In addition to broadening, collisions induce a shift of the absorption line's central position $v_{0,i,j}$ which is likewise separated into impacts by self- and foreign species. The disturbed central line position $v_{i,j}$ calculates to:

$$v_{i,j} = v_{0,i,j} + \delta_{i,j,s} p_s + \sum_f \delta_{i,j,f} p_f \qquad \text{(Eqn. 36)}$$

with the self- and foreign-induced pressure shift coefficients $\delta_{i,j,s}$ and $\delta_{i,j,f}$, respectively.

The pressure and temperature broadening as well as the shift coefficients can be calculated or measured for each target molecule, transition and foreign molecule. For a comprehensive calculation of even a single line, this results in a large number of coefficients.

In addition to the natural line width, the Doppler and collisional broadenings, further effects are known to have an impact on the absorption line's shape, such as the Dicke narrowing [91], saturation broadening and time-of-flight broadening. However, these effects are negligible in typical combustion environments.

Line Shape

The intensity of a light beam which is guided through a gas sample is attenuated due to absorption by the transitions. Absorption spectra are usually observed within units of the optical density $OD_e(v)$, which compares the attenuated to the original light intensity

$$OD_e(v) = -ln\left(\frac{I}{I_0}\right)(v) = \alpha(v,T) \cdot L \qquad \text{(Eqn. 37)}$$

with the absorption coefficient α and the absorption length L, which is the propagation distance of the beam within the sample. The absorption coefficient for a single transition i of molecule j is related to the absorption line strength and the number density of molecules within the gas sample n_j via

$$\alpha_{i,j}(v,T) = S_{i,j}(T) \cdot \phi_{i,j}(v,T) \cdot n_j \qquad \text{(Eqn. 38)}$$

Here, the absorption line-shape is separated from its intensity for a more convenient representation. Therefore, the line-shape function $\phi_{i,j}$ has to be area-normalized with $\int \phi_{i,j}(v,T)\, dv = 1$. As discussed above, the absorption lines can either be described by a Gaussian line-shape function if Doppler broadening is dominant, or by a Lorentz profile for a stronger collisional broadening. Within most combustion environments, the Doppler and collisional broadening widths are in the same order of magnitude. The convolution of both profiles is the Voigt profile, which does not have an analytical solution. However, it can be gained via approximate methods [92,93] or via tabulated functions [94]. In this work, the approximate method by Humlíček [93] is used, with a Voigt-width Δv_V of

$$\Delta v_V = 0.5346 \cdot \Delta v_C + \sqrt{0.2166 \cdot \Delta v_C{}^2 + \Delta v_D{}^2} \qquad \text{(Eqn. 39)}$$

In fig. 2.12, the Lorentz, Gauss and convoluted Voigt profiles are compared for identical widths. The Lorentz-profile is decaying more slowly at the outer flanks, whereas the Gauss is stronger close to the central wavelength.

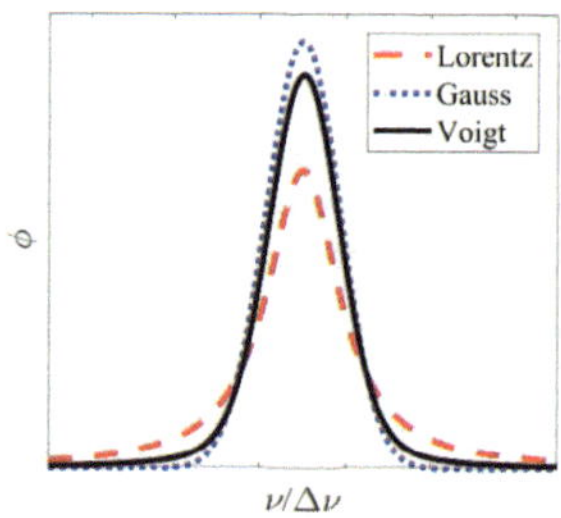

Fig. 2.15: Comparison between a Lorentz, Gauss and convoluted Voigt profile of identical width.

The Voigt profile is the most commonly used description of the absorption line profile, as it can relatively easily be calculated. However, if Dicke narrowing is considered, Galatry or Rautian profiles are more appropriate depending on the model of the collision (hard or soft model). Highly accurate profile measurements allow to detect deviations from the model profiles even beyond the Galatry and Rautian profiles to be detected. To compensate for this effect, a multitude of different profile functions is known for modeling the different physical aspects. Further details on these models can be found in

[95,96]. However, as typical signal-to-noise ratios achieved in combustion absorption measurements are too low to resolve differences in the models, Voigt profiles are used in this work.

2.2.3 Spectroscopic principle: TDLAS

By measuring the intensity and shape of absorption lines, the absorption spectroscopy utilized in this work allows to determine the gas temperature and species concentrations within combustion environments. One prominent implementation of absorption spectroscopy which allows high spectral resolutions is the tunable diode laser absorption spectroscopy (TDLAS). This technique has been applied to study a multitude of applications [97–114], as it is contactless, calibration-free and, compared to competitive techniques, provides a high measurement repetition rate (~kHz–MHz). In addition, TDLAS yields quantitative results, which is advantageous for the interpretation of the phenomena under investigation. The necessary optical components are comparatively cheap and can be set up to be fiber-coupled. This allows to easily separate the spectrometers from the location of the measurement object in large-scale applications without the need for a precise laser-alignment, and to operate the spectrometers in harsh environments in terms of temperature, mechanical stress or dust. In addition, fiber-coupled devices rely on significantly smaller optical elements than free-beam setups, which allows to construct minimally-invasive probes. Goldenstein et al. recently presented an excellent review of the technique and the application to combustion research [96]. However, TDLAS has rarely been used in oxy-fuel environments [52].

In the TDLAS technique, light emitted by a narrowband laser is guided through the gas sample (e.g. the combustor), while the wavelength of the laser is quickly tuned across a spectral region. Due to resonant and non-resonant absorption processes, part of the light is absorbed and the intensity attenuated. Assuming a homogeneous absorption path, this process can be described by eqn. 37, which is known as the Beer-Lambert law. Combined with eqn. 38, it takes the form:

$$OD_e(v) = -ln\left(\frac{I + \Theta(t)}{I_0 \cdot Tr(t)}\right)(v) = \sum_{i,j} S_{i,j}(T) \cdot n_j \cdot \phi\left(v - v_{0,i,j}, T, p_j\right) \cdot L \qquad \text{(Eqn. 40)}$$

where $I_0(v)$ and $I(v)$ define the incident and the attenuated laser light intensity, respectively. The optical density is described by the temperature dependent line-strength $S_{i,j}(T)$ of transition i of the species j at the center frequency $v_{0,i,j}$, the partial pressure p_j and the number density n_j of species j. L depicts the absorption length.

In combustion environments, background emission (e.g. chemiluminescence or thermal radiation of hot surfaces) and non-resonant absorption (e.g. by particles or windows) pose additional factors to the signal. As the maximum of the blackbody radiation at combustion-relevant temperatures (1000 – 2500 K) falls within the near-infrared region (NIR), the detectors in this work are sensitive to these emissions which can substantially exceed the laser intensity. Large particles or partially-transparent windows cause an intensity attenuation through absorption. Particles with sizes in the order of the wavelength (~1–10 μm), however, provoke Mie-scattering of the laser beam, which reduces the measured laser intensity further due to partial scattering. For high particle densities in coal combustion, the attenuation has been shown to be substantial [115]. Temperature gradients in the gas sample lead to density and thus refractive index variations. These variations cause beam-steering, which also reduces the measured laser intensity through misalignment. In eqn. 40 these effects are accounted for through the extension terms $Tr(t)$ for non-resonant absorption and $\Theta(t)$ for background radiation.

Fig. 2.16a shows the influences of the different factors on the TDLAS-signal. A linear increase of the laser injection current results in a non-linear laser baseline I_0 (1 in fig. 2.16a), which is partially absorbed at the wavelength of a molecular transition (2). Broadband attenuation is reducing the laser intensity by a factor $Tr(t)$, which needs to be constant over the duration of one scan (3). As the laser baseline is typically identified by a polynomial fit to the signal, the broadband attenuation only changes the slope of the baseline and can easily be separated. Background emissions add to the signal by means of an offset (4) which can be determined by reducing the laser power below the threshold current prior to a scan. The combined effects lead to a measured intensity I (5).

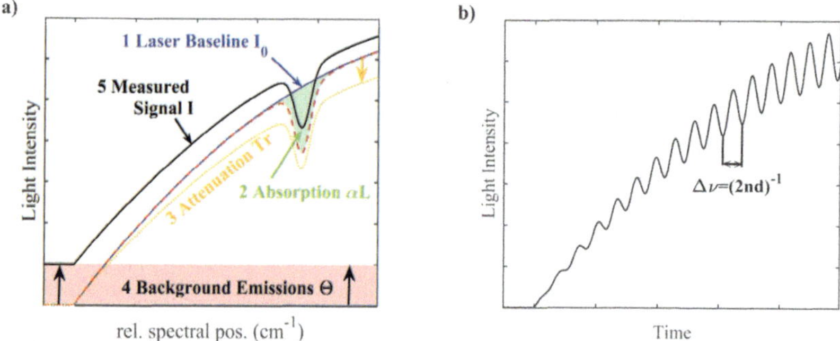

Fig. 2.16: Working principle of TDLAS for a single absorption line and factors in the Beer-Lambert law; b) Airy-function on laser baseline due to propagation through an etalon.

Rearranging eqn. 40 with respect to n_j for a specifically-measured transition i, integrating the line shape function over the spectral tuning range and applying the ideal gas law (eqn. 7) yields the absolute gas concentration x of species j

$$x_j = -\frac{k_B \cdot T}{S_i(T) \cdot p \cdot L} \cdot \underbrace{\int \ln\left(\frac{I(\nu) - \Theta(t)}{I_0(\nu) \cdot Tr(t)}\frac{\partial \nu}{\partial t}\right) dt}_{:= A}$$

(Eqn. 41)

The tuning function $\partial\nu(t)/\partial t$ allows a transformation of the measured signal from a temporal to a spectral representation and is a characteristic parameter of the diode laser. It can be measured by using a planar Fabry-Pérot etalon in the laser path, which provokes interference patterns of the intensity due to internal reflections. Constructive interference occurs for an optical path length within the etalon of multiples of the laser wavelength. If this criterion is not met, negative interference provokes a baseline modulation by means of an Airy-function (fig. 2.16b). The spectral distance between two maxima is given by the free spectral range of the etalon

$$FSR = \Delta\nu = \frac{1}{2n_{ref}d}$$

(Eqn. 42)

with the length of the etalon d and the group index n_{ref}. In this work, sapphire etalons of various thicknesses were utilized to measure the tuning function. As diode lasers operate highly reproducibly, measuring the tuning function once for every operating condition is sufficient.

The temperature-dependent line strength, as a consequence of the Boltzmann-distribution (eqn. 28), is a characteristic of each transition. While the line strengths of transitions with high lower state energy E_i are increasing with temperature, they are decreasing for those with low E_i. Hence, the ratio of the measured absorption area A_j [21] of two absorption lines with different lower state energies are an unambiguous function of the temperature. This allows a determination of the path-averaged gas temperature T through TDLAS using two-line thermometry. Assuming homogeneous conditions and utilizing eqn. 31, the absorption area ratio of two adjacent[22] lines 1 and 2 of the same species calculates to:

---

[21] The line area A_j of transition j is defined as the integral term in eqn. 41.

[22] To simplify eqn. 31, $(\nu_{0,1} - \nu_{0,2})/\nu_{0,1} \approx 0$ is required.

$$\frac{A_1}{A_2} = \frac{S_1(T_{296})}{S_2(T_{296})} \cdot e^{-\frac{(E_{i1}-E_{i2})}{k_b}\cdot\left(\frac{1}{T}-\frac{1}{T_{296}}\right)} . \qquad\text{(Eqn. 43)}$$

Hereby, $S_j(T_{296})$ is the line strength of the transitions at the reference temperature 296 K.

The detection limits of the direct TDLAS described here can be improved by using wavelength modulation absorption spectroscopy (WMS). In this technique, the laser wavelength is not only modulated in a linear manner but with a superimposed sinusoid of frequency f. The interaction between the rapidly modulating wavelength and the nonlinear absorption features allows to isolate harmonic components in the signal. Lock-in amplifiers at the second harmonic ($2f$) allow a separation of the absorption and a determination of the concentration [116]. Using the procedure of normalizing the $2f$ signal with the $1f$ signal recently published by Goldenstein et al. [117] allows quantitative measurements without a need for calibration, similar to direct TDLAS. Even though the technique rejects low-frequency noise through filtering and thus reduces the uncertainties, it is not employed in this work, as the experimental and post-processing efforts are unnecessarily increased. As will be seen later, the observed standard deviations were dominated by process fluctuations rather than measurement uncertainties in the vast majority of cases.

2.2.4 Allan-Werle Variance

Spectra obtained by TDLAS are subject to various sources of errors, such as detector electrical white noise (f^0-noise), laser noise (f^{-1}-noise) or drifts (f^{-2}-noise)[23]. Therefore, averaging of the spectra over multiple scans is often required. However, while this procedure initially reduces the electrical noise to a certain extent, drift effects distort the spectra for a large number of scans. Hence, an optimal number of scans for averaging needs to be found.

The Allan variance is a tool to study the frequency-stability of measurement systems, such as clocks, oscillators and gyroscopes. It yields the optimal number of samples for averaging to minimize different sources of error. Werle et al. [47] first described the application of this procedure to a TDLAS system in order to achieve a minimal detection limit, where it is renamed to Allan-Werle variance (AWV). To obtain the AWV, a time series of concentration measurements $\chi_{1...,N}$ with a large number of scans[24] is recorded for a constant measurement frequency f first and then processed without averaging. The N elements of the original data set can be divided into m subsets with corresponding subensemble numbers $s = 1...,m$. Each of the subensembles contains $k = N/m$ elements and requires a duration of $\tau(k) = k/f$ to be measured. For each of these m subgroups an average value $A_{s=1...,m}(k)$ can be calculated:

$$A_s(k) = \frac{1}{k} \cdot \textstyle\sum_{l=1}^{k} \chi_{(s-1)k+l}. \qquad\text{(Eqn. 44)}$$

The Allan-Werle variance $\sigma^2{}_{AWV}$ of the sample size k is then defined as:

$$\sigma^2{}_{AWV}(n) = \frac{1}{2m} \sum_{s=1}^{m-1} [A_{s+1}(k) - A_s(k)]^2 \qquad\text{(Eqn. 45)}$$

.

---

[23] White noise is constant in the frequency space, hence it is referred to as f^0-noise. The other sources of noise are linear with a slope of -1 (f^{-1} "pink" noise), -2 (f^{-2} "brown" noise) and -3 (f^{-3}-noise) in the double-logarithmic power spectral density.
[24] Practice has proven $N{\sim}10^4{-}10^5$ to be sufficient for most combustion-related TDLAS systems, which exceeds the values given by Werle et al. [118] for environmental applications by one order of magnitude.

The Allan-Werle variance is commonly displayed as a function of the averaging sizes k or the corresponding averaging time $\tau = k/f$ in a double-logarithmic plot (fig. 2.17). In this representation, the AWV has a global minimum. For a corresponding averaging size k or averaging time τ, the effects related to noise are minimized and a minimal detection limit and a maximal signal-to-noise ratio is obtained. However, this requires the effective measurement frequency to be reduced by $f' = f/k$.

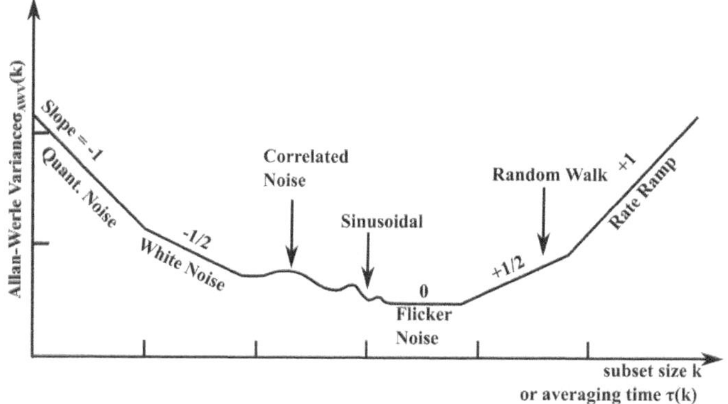

Fig. 2.17: Sample plot of Allan-Werle variance for complex systems with various noise effects [119].

For stable measurement environments, such as reference cells or environmental applications [118,120–122], the Allan-Werle variance is initially decreasing due to reduced impact of white noise with increasing averaging sizes. The slope of this decrease is 1/2. For very high averaging sizes, the variance is finally increasing (right side of fig. 2.17) due to drifts within, e.g., the lasers. While these two effects are the only ones occurring during TDLAS measurements of stationary measurement conditions, more complex systems (e.g. gyroscopes) are generally known to have an AWV influenced by other effects as well (fig. 2.17) [119]. These effects include quantum noise, sinusoidal signals, random walk of the signal and non-randomly distributed fluctuations, which all occur at different regions and with different shape in the AWV.

2.2.5 Diode Lasers

In this work, narrow-band diode lasers are employed to generate light which is quickly tuned across entire absorption features. Due to their widespread utilization in telecommunication and sensor applications, these lasers are comparatively cheap and their technology is mature. Diode-lasers are based on semiconductor alloys, which are p- and n-doped with elements of the main groups III and IV. If a current is applied to the semiconductor in forward direction, holes of the p-doped and electrons from the n-doped material diffuse towards the semiconductor junction, where they recombine and emit the energetic difference between the conduction and valence band as radiation (E_g in fig. 2.18a and b). This process generates a population inversion of the energetic states in the junction, which is a prerequisite for lasing. Without further measures, spontaneous emissions from the upper state are dominant, and the semiconductor operates as LED. Applying an optical resonator to the active material selectively amplifies certain modes within the gain profile for induced emission (fig. 2.18 c). The spectral location of this profile is defined by the band gap of the utilized materials, and its shape by the temperature-dependent population of states in the valence-and conduction band. Hence, through variations in the doping and semiconductor materials, lasers can be set up in a wide range of emission wavelengths. Diode lasers in near infrared (NIR) and visible regions below 1300 nm are typically grown on InP substrates and doped with Al, Ga, As and P compounds. For higher wavelengths up to 3 µm, the lasers are grown in GaAs and In, Sb and P are employed as alloys. These materials typically have a lifetime of the excited state in the magnitude of 10^{-13} s and a gain width in the order of 40 nm, which requires a narrowing for the induced emission to be dominant over spontaneous emissions.

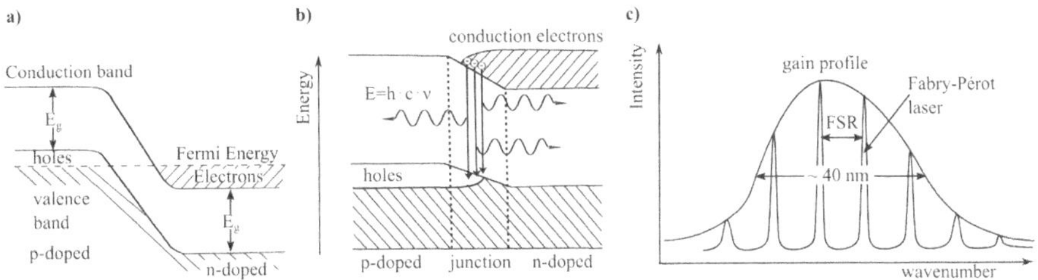

Fig. 2.18: Semiconductor working principle: a) Interface without external potential [87]; b) externally applied forward current [87]; c) semiconductor gain profile and Fabry-Pérot laser modes.

The simplest realization of a diode laser is the Fabry-Pérot type, where the front and rear surfaces of the semiconductor material are polished. The partial reflections at the interface generate an internal resonator whose free spectral range (eqn. 42) is high due to the short dimensions of the laser (resonator length ~0.5 mm). However, as multiple possible modes are located within the gain spectrum, the laser is multi-modal and allows random mode-hopping with sudden changes of the emission wavelength. These characteristics impede a utilization of Fabry-Pérot lasers for TDLAS, which require a mode selection.

In this work, three different types of single-mode diode lasers are utilized which are all based on diode-lasers but employ different mode-selection features: the distributed-feedback laser (DFB), the discrete mode laser (DM) and the vertical-cavity surface-emitting laser (VCSEL). In each of the laser types, typically five pn-junctions are stacked atop each other to form a multiple quantum well heterostructure, which is necessary to increase the laser intensity within the junctions and thus reduce the lasing threshold current. In addition, the modulations of the refractive index act as a waveguide.

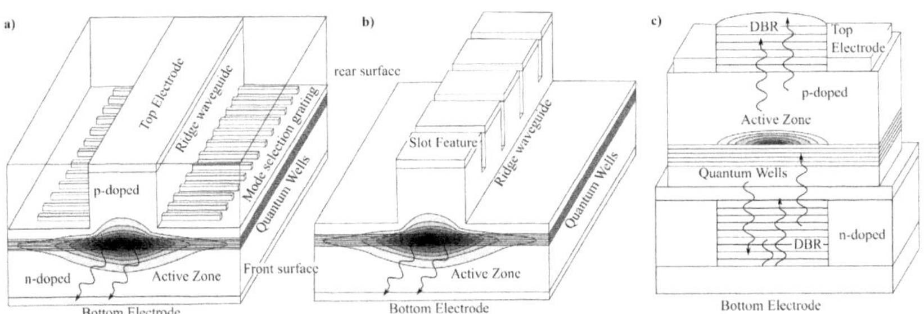

Fig. 2.19: Structure of single-mode diode lasers: a) Distributed feedback laser (DFB); b) Discrete mode laser (DM); c) Vertical cavity surface emitting laser (VCSEL).

The distributed feedback laser (fig. 2.19a) is edge-emitting, as the resonator surfaces are orthogonal to the active gain layers. Hence, the laser emission profile is roughly elliptical. The p-doped upper end of the laser is shaped to a ridge to guide the photons along the resonator axis. Etched metal gratings along this ridge waveguide pose a mode selection to the emission spectrum. A part of the wave within the ridge is coupled into the grating by means of an evanescent wave. All modes whose multiple of the wavelength does not match the optical spacing of the grid are diffracted and thus suppressed, leaving a single, dominant mode whose width is determined by the grating. The side-mode suppression ratios of these lasers are in the order of 30–60 dB, which is sufficient for TDLAS-applications.

Besides this loss-grid type, a mode selection in DFBs can also be performed via a periodic index modulation in the ridge waveguide (distributed Bragg reflector, DBR), which corresponds to an interference filter. These lasers are similar to distributed Bragg grating lasers, where the DBR is placed before and behind the ridge waveguide. Strictly speaking, a DFB does not have a real resonator, as the emitting surface is coated with anti-reflective layers. These layers are necessary, as during the cleavage process of the lasers, the length of the cavity cannot be manufactured precisely, and the corresponding Fabry-Pérot interferences modes would compete with the grating modes. For this reason, the laser spectral bandwidth of this laser type is comparably broad (10 MHz) and optical feedback into the laser by external optical elements leads to mode-hopping.

The second laser type, the novel discrete mode laser [123,124], avoids this drawback (fig. 2.19b). The structure of these lasers is very similar to DFBs. However, instead of a metal grating, the mode selection is performed by slot features within the ridge waveguide, which creates a superposition of interference patterns. For these lasers, none of the surfaces are anti-reflective, which introduces a primary mode selection by means of Fabry-Pérot interference. Reflections between the slot patterns, which have a typical spacing around 6 μm, introduce additional Fabry-Pérot patterns with a higher finesse, as this cavity is shorter. The superposition of these patterns leave only a primary mode, which has a similar side-mode suppression ratio as the DFBs. Due to the higher finesse of the mode selection, however, the lasers are reported to have a significantly smaller spectral bandwidth (~500 kHz) [125]. In addition, the mode selection technique causes an inherent insensitivity to optical feedback, which even intensifies the mode selection. To the best of the author's knowledge, these lasers have not been reported for combustion-related TDLAS measurements before the author's work.

In contrast to the edge-emitters, the resonator surfaces of VCSELs are in parallel to the quantum well heterostructure (fig. 2.19c) which introduces a circular mode field. Even though this allows the laser light to be coupled into optical fibers more easily, the output power of these lasers is significantly lower than for edge-emitters (<1 mW compared to ~20 mW of edge-emitters), as the active zone is thinner. The mode selection of these lasers is performed via distributed Bragg gratings atop and below the quantum well heterostructure, which cause a broader spectral bandwidth in the order of ~100 MHz. However, the smaller dimensions of the laser make it favorable for a lot of applications, as the smaller thermal capacity allows significantly higher modulation speeds and a broader spectral tunability. In addition, more lasers fit on a Ga-waver, which reduces the laser price.

In all three laser types, the output wavelength is coarsely defined by the gain spectrum of the used materials. A finer tuning of the emission wavelength can be performed by the laser temperature. Increasing the laser temperature lengthens the optical cavity due to thermal expansion and reduces the band gap between the valence band and the conduction band due to changes in the population of states. Both effects reduce the emission wavelength of the dominant mode. Hence, the lasers are equipped with Peltier elements for temperature control, which allows to stabilize or tune the wavelength by a few nanometers. However, due to thermal inertia of the laser, the tuning is rather slow (~mHz-Hz). A rapid tuning of the wavelength can be performed through a modulation of the injection current. An increase of the current leads to higher ohmic losses and thus temperatures in the quantum well heterostructure, which leads to effects similar to the temperature tuning. However, as the thermal inertia of the heterostructure is smaller, the modulation bandwidth is higher than via direct temperature tuning. In addition, an increased current leads to a higher carrier density, which in turn changes the refractive index and thus the emission wavelength. Due to the nonlinear dependency of these effects on the current, the tunability is higher for high currents than those in the order of the threshold current.

Similarly to the temperature tuning, an increased injection current increases the emission wavelength. The small dimensions and thus thermal inertia of the VCSELs allow modulation frequencies in the magnitude of 100 MHz without a significant reduction of the tunability, which is around 20 cm^{-1}. Edge emitters, however, can only be modulated up to 10 kHz and have a spectral tunability of ~2-5 cm^{-1}. Hence, VCSELs are favorable where a broad spectral range needs to be covered, e.g. where multiple lines overlap or are spectrally broadened due to high pressures. More information on diode lasers can be found in [87,126–132].

2.3 Residence Time Distributions

Oxy-fuel combustion is mainly affected by the flow, the chemistry and the thermodynamics of the process. While TDLAS allows direct measurements of temperature and species concentrations, which are relevant for the latter two, no information is gained on the flow field. Typically, local flow field characteristics such as velocity, strain rate, vorticity and slip velocity in two-phase flows are investigated in combustion research. The most common techniques, particle-image velocimetry and laser-Doppler velocimetry, allow local 0D, 2D and even tomographic 3D measurements of the velocity at high measurement rates. However, less attention has been paid to global quantities of the flow.

The gas residence time, which is the mean time a fluid element remains inside the combustor, is of specific interest for combustion reactor modeling [133–137]. It strongly affects the chemical reaction progress and therefore the flue gas composition and the fuel burnout rate. In solid fuel combustion processes, the devolatilization and the pyrolysis are depending on the particle residence time. In addition, the residence time is a valuable parameter for experimental validation of numerical flow simulations using large-eddy (LES) or Reynolds-Averaged-Navier-Stokes simulations (RANS). As individual fluid elements in non-ideal flow systems travel along different streak lines from the inlet to the outlet and are exposed to different velocities, the elements exhibit different residence times, such that a residence time distribution (RTD) is assessed. In turbulent systems, the residence time is additionally non-stationary, as, due to the turbulent features in the flow, the RTD-curve is randomly altered over time. Hence, a mean RTD-curve cannot be measured by a single measurement and as such standard deviations need to be quantified.

Due to high process temperatures, experimental determinations of the RTD within combustors are sparse, difficult to achieve and exhibit large systematical errors. The aim of this section is to first give an introduction into the fundamental strategies of measuring the RTD, independently of specific realizations or measurement techniques. The strategies are discussed with respect to their advantages and disadvantages. In addition, a brief overview is given on RTD modeling in generic setups. Secondly, an overview of the technical realizations of RTD measurements reported in literature is given and their individual strengths are evaluated.

2.3.1 Measurement Principle

Generally, the residence time distribution is accessed via injection of a non-reacting, chemically inert tracer into the inlet flow and measuring the system response by means of the tracer concentration in the exiting stream over time. The injection can be achieved through one of two canonical injection schemes [138]:

- *Pulse injection:* The tracer is injected into the inlet flow at t=0 for an ideally infinitely short duration (delta-peak $\delta(t)$). In practice, the duration of this pulse needs to be significantly shorter than the mean residence time. The area-normalized tracer concentration curve at the outlet directly yields the RTD $E(t)$ (fig. 2.20a). To avoid a significant influence on the flame chemistry or flow, the tracer peak concentration during the injection must be negligible low (typically around 1 % of the inlet flow or less). As the pulse is stretched in the reactor, this leads to significantly lower concentrations at the outlet. This effect is further enhanced in combustors where typically only one of the inlet flows, which is diluted by supplementary streams, is traced. Hence, this technique is rarely employed at combustion chamber experiments.

- *Step injection:* The injection of the tracer into the inlet flow starts at t=0 by means of a sudden increase but continues afterwards. The tracer molecules begin to accumulate within the reactor. The response of this excitation is the cumulative distribution function (CDF) $F(t)$, which can be measured by normalizing the outlet temporal evolution of the concentration with the steady state concentration $F(t) = \chi(t)/\chi_\infty$ (fig. 2.20b). The residence time distribution and the CDF are connected by:

$$E(t)=dF(t)/dt$$

As the tracer molecules accumulate within the reactor, the outlet concentrations significantly exceed those of an impulse injection in most applications. For a single-inflow single-outlet reactor with constant number of molecules, the outlet steady state concentration is identical to the concentration at the inlet. Even though the tracer concentrations at the outlet are higher compared to the aforementioned scheme, this technique requires differentiation of the CDF to obtain the RTD, which is prone to errors due to amplification of measurement noise. The step input can be either positive by a sudden start of a tracer injection ("step up") or negative by a sudden stop of the injection ("step down") which is followed by a washout. The normalized concentration of the latter yields the washout function $W(t)$, which is connected to the CDF by simply $F(t) = 1 - W(t)$. The washout function is often reported to be less dependent on turbulence [139] [E1], even though the origin of this behavior is not clear.

Besides the two injection strategies, the RTD can also be measured by e.g. sinusoidal or triangular excitation. These functions, however, can mathematically be decomposed into a superposition of multiple step and impulse injections [138] and are therefore not considered any further.

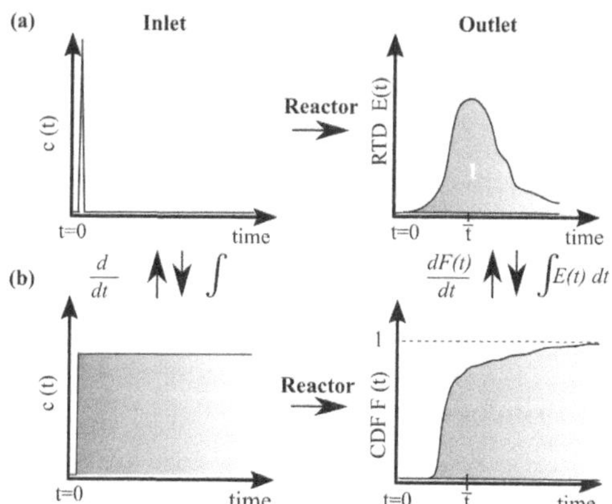

Fig. 2.20: Tracer injection schemes and conversions: Pulse injection (a), positive step injection (b).

From the RTD, the mean residence time of fluid elements within the reactor $\bar{t}$ and the higher statistical quantities can be obtained [140]:

$$\bar{t} = \int_0^\infty t \cdot E(t)\, dt \tag{Eqn. 46}$$

$$\sigma_{RTD}^2 = \int_0^\infty (t - \bar{t})^2 \cdot E(t)\, dt \tag{Eqn. 47}$$

$$S^3 = \int_0^\infty (t - \bar{t})^3 \cdot E(t)\, dt / \sigma^3 \tag{Eqn. 48}$$

The residence time variance σ_{RTD} describes the dispersion of the RTD in terms of the moment. The skewness is a measure for the deviation from a symmetrical distribution ($S = 0$). A left-skew distribution exists for S < 0 and a right-skew distribution for S > 0. Even though the CDF can be compared to numerical simulations, it does not yield the higher statistical moments without differentiation. In this respect, the RTD is more enlightening.

However, based on the CDF the Bodenstein-number (Bo) can be calculated, which describes backmixing of the reactor through the ratio of the convection flow to the axial diffusion flow: $Bo = uL/D_{ax}$, with D_{ax} the axial diffusion coefficient, L the characteristic length of the reactor and u the convective flow velocity. The Bodenstein-number is a flow analogy to the Peclét-number $Pe = LU/\alpha$ and the Reynolds-number Re, which describe the ratio of convective to the diffusive heat flow and impulse flow, respectively.

For $Bo \rightarrow \infty$ the axial diffusion is negligible and the reactor is described by a plug flow, whereas for $Bo \rightarrow 0$ the flow is ideally mixed leading to a well stirred reactor. According to Hagen [138] the Bodenstein-number can be estimated from the slope of the CDF at the mean residence time:

$$Bo = 4\pi \cdot \left(\left[\frac{dF(t)}{d(t/\bar{t})} \right]_{t/\bar{t}=1} \right)^2 \qquad \text{(Eqn. 49)}$$

Based on CDF or RTD measurements, real reactors can be modeled as a network of one or multiple plug flow reactors (PFR) and well stirred reactors (WSR) in series or parallel. In these networks, the PFRs account for a delay in both the CDF and the RTD, which can be described by the Heaviside function $H(t - \bar{t}_{PFR})$ in the CDF and a delta function $\delta(t - \bar{t}_{PFR})$ in the RTD (fig. 2.21). Hereby, $\bar{t}_{PFR}$ is the delay time of the plug flow. A WSR exhibits an exponentially decaying RTD, which can be accounted for by multiplying a factor of $F(t) \propto 1 - exp(-t/\bar{t}_{WSR})$ in the CDF and $E(t) \propto exp(-t/\bar{t}_{WSR})$ in the RTD, with $\bar{t}_{WSR}$ the convective decay constant of the recycle flow.

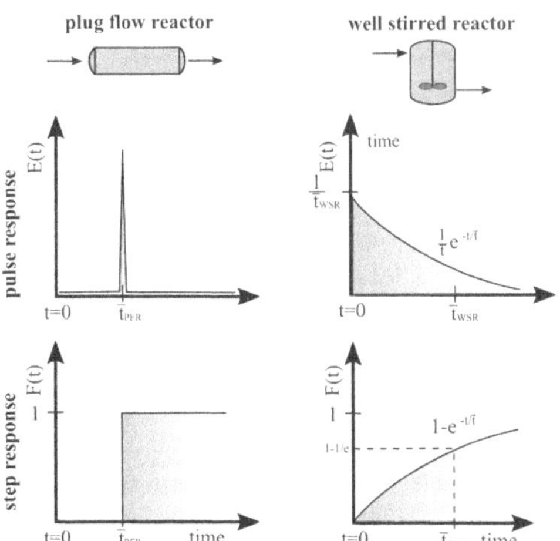

Fig. 2.21: RTD and CDF for the idealized reactor models of a plug flow or a well stirred reactor.

Further information on principle considerations to residence time and reactor modeling can be found in [138,141].

2.3.2 State of Research

In most combustion related literature, the mean residence time is estimated by comparing the volume of the reactor V_R to the inlet volume flow rate $\dot{V}$ [142]: $\bar{t} = V_R/\dot{V}$. While this procedure may be sufficient for ideal, non-reactive flow conditions without recirculation or axial diffusive transport, e.g. plug-flows in pipes, it fails under turbulent, reactive multi-dimensional flows. For these processes, measurements or simulations are required.

To reduce the experimental complexity, combustor flow properties, such as the RTD, are commonly mimicked by using non-reacting gaseous or liquid fluids only [137,143–145]. In these previous studies, RTDs are accessed by measuring the concentration of a dye or gaseous tracer using either suction probes in combination with gas analyzers or laser induced fluorescence. However, even if the Reynolds-numbers of the inflows are kept identical to reacting conditions, the non-reacting operation fails to simulate the effects of expansion and temperature increase which alter the density, velocity, viscosity and chemical composition. Even under these simplifications, precise measurements of the RTD are complex.

Van der Lans et al. [144], for example, accessed the RTD via an injection of He into a gaseous non-reactive airflow of a model combustor. A hot wire suction probe was placed in the outlet flow, quantitatively measuring the He-concentration through the significant difference in heat conductivity between air and helium. The authors recognized the tactile probe to disturb the flow by its physical presence and by extracting gas. The tracer was to be fed into the chamber by means of a negative step injection to perform washout experiments with detectable He-concentrations. The authors had to apply a severe "all in window" filter to reduce the effects of noise, which were amplified through the derivation, and thus had to reduce the temporal resolution to 1 s. The low temporal resolution and the measurement technique itself impede a utilization for reactive high-temperature processes, specifically with particles in the flow.

In-situ measurements of the RTD of gases in combustors under high temperature and chemically reacting conditions are sparse. They can be achieved by injection of solid, liquid or gaseous tracers which are then detected by appropriate diagnostics within the processes. Rao et al. [146] obtained the RTD in a gasifier by injection of radioactive gaseous tracers into the combustion process and consecutive measurements at different locations using scintillators. Costs for tracers and equipment, their environmental impact and inability to distinguish between signals from different but adjacent positions make this technique impractical for laboratory measurements. Göckeler et al. [143] measured the density of seeding particles after an injection into a premixed combustor using laser-sheet illumination and high-speed cameras. As the tracer particles can hardly be distinguished from fuel particles, this technique is not applicable to solid-fuel combustors. Moreover, the injection of solid seeding particle by either a pulse or a step temporal profile seems challenging. In addition, the outlet signal is very noisy, which impedes to deduct higher statistical moments or derivatives without filtering. Therefore, a detection of gaseous tracers is preferable.

Schultz et al. [108] used absorption spectroscopy to estimate an upper bound of the residence time in a scramjet combustor by performing flame extinction experiments and observing the reducing concentrations of combustion products. However, flame extinction experiments are not capable of resolving residence time distributions, as the flow field, chemistry and density are changing during the event of an extinction. Schlosser et al. [147] applied TDLAS to measure RTDs in a waste incinerator using gaseous alkaline atoms as tracers. Rubidium-salt was dispersed in a liquid that was injected into the combustion process. Hereafter, the liquid evaporated prior to sublimation of the salt. These complex phase transitions impede the use of these salt solutions as tracers in lab scale combustors, as their impact on the measured RTD for short residence times remains unclear. However, the technique is promising for measurements in particle-laden reactive flows.

3 Spectrometer Design

"Without measuring, science is just philosophy"

- Inspired by an engineering proverb

Oxy-fuel combustion is influenced by a mutual dependence of flow, chemistry and thermodynamics. For a detailed experimental investigation of the processes involved, relevant quantities of these fields have to be identified and suitable techniques to measure these quantities to be chosen.

While comprehensive investigations of the complex flows in oxy-fuel combustion are available [23,37,38,53], less attention is payed to the thermochemical state of the combustion, which is relevant for both chemistry and thermodynamics. The two properties of the gas-phase's thermochemical state – temperature and species concentrations – are highly relevant for combustion systems through their impact on reaction rates, heat transfer through diffusion or radiation, reaction progress and the physical properties of the gas-phase, such as density, heat capacity, diffusivity and viscosity. However, most conventional temperature and species concentration measurement techniques fail under operation of the high temperature environments of oxy-fuel combustion, which include high radiation levels, eventually high particle loads and ongoing chemical reactions. In terms of temperature measurement, conventional tactile methods, such as thermocouples, have to be radiation-corrected. This process is prone to large errors, due to blackbody radiation from hot combustion chamber walls and the intense radiation from the combustion gases which are particularly enhanced during oxy-fuel combustion through the presence of high CO_2 and H_2O concentrations. Suction-pyrometry overcomes this issue; however, the bulky probes necessary for this technique cannot be used in small burners and are difficult to operate even in large-scale applications. For these reasons, contact-free optical measurement techniques are preferred in combustion research.

The need for a universal applicability to oxy-fuel and oxy-gas combustion rules out most popular optical gas temperature measurement techniques, as the presence of particles impedes measurements by Rayleigh-scattering [148–150], laser induced grating spectroscopy (LIGS) [151], phosphor-thermometry [152,153] and laser induced fluorescence (LIF)–thermometry [154–156]. Additionally, without knowledge of the species mole fractions, the Rayleigh-cross-sections and LIF-quenching coefficients are unknown, which impede these measurement techniques in oxy-fuel processes. Laser induced breakdown spectroscopy (LIBS) has been demonstrated to be capable of measuring the gas–phase temperature in solid-fuel combustion environments, but is not exactly minimally invasive, due to the ignition of a thermal plasma [157].

Coherent Anti-Stokes Raman (CARS) [33] is technically feasible in oxy-fuel combustion, however, most evaluation codes rely on spectra of nitrogen for the CARS-fit, which is not present in oxy-fuel combustion. Additionally, in flows with high particle loads, the CARS signal is accompanied by resonant and non-resonant laser-induced plasma processes on particles by the pump and probe lasers, which makes a signal separation complex [158]. In addition, all aforementioned methods rely on high-power lasers, which are comparatively expensive, bulky and difficult to operate outside of controlled environments, e.g. in large-scale combustion facilities.

Similar considerations as for the temperature measurements hold for the most common species concentration measurement techniques, which include Raman spectroscopy [18,22,149,150,159], CARS, LIBS and LIF. Ex-situ techniques, such as FTIR-spectroscopy [31], gas-chromatography [160], mass-spectrometry [161] or conventional gas analyzers based on chemical absorption of the gases, UV photometry and nondispersive infrared absorption [19,32,54] rely on suction of combustion gases out of the combustor. Through this approach, short-living radicals and species are eliminated and cannot be measured. Additionally, as these methods are rather slow, the information on the temporal fluctuations is lost which is particularly obstructive in turbulent combustion research.

Absorption spectroscopy methods, in particular tunable diode laser absorption spectroscopy (TDLAS), have proven their versatile usability in combustion systems. As described in sec. 2.2, the technique has been reported as being particularly capable of measuring the gas temperature and/or the concentration of individual species. The gas temperature is often measured through two-line thermometry by taking the absorption line area ratio of two spectrally adjacent absorption lines of the same species, which is temperature dependent. However, TDLAS-systems for combustion research are not commercially available and need to be designed application-dependently. For each investigated species, at least one suitable absorption line with low uncertainties in the line coefficients needs to be identified. Only few absorption lines are available in databases with a full set of high quality line coefficients. Hence, where necessary, the line coefficient quality of the chosen absorption lines needs to be improved through experiments or literature review. Additionally, the technique was mostly used to measure the concentration of one or a few combustion species alone. For a quasi-simultaneous measurement of multiple thermochemical quantities at the same location, the technique needs to be extended.

These needs define the aims of this chapter:

- selection of suitable absorption lines to measure the chemical composition of multiple combustion relevant species in a wide range of thermochemical states using the TDLAS technique,

- selection of a set of absorption lines to measure the gas temperature based on absorption spectroscopy,

- development of a technique to measure the aforementioned quantities quasi-simultaneously,

- to identify suitable lasers to acquire the absorption spectra experimentally,

- development of a versatile optical setup that can be used in a wide parameter range of burners, temperatures and chemical compositions and is easily scalable, robust and affordable,

- to improve the line coefficient quality of the selected absorption lines in databases where necessary, either through literature research or new experimental investigations,

- to develop an experimental setup and measurement strategy to measure the line coefficients,

- to accurately measure the necessary line coefficients and comprehensively analyze the measurement uncertainties.

The process of the absorption line selection is described first, followed by the results of this analysis. The corresponding laser setup, the data-acquisition system and general post-processing steps are described hereinafter together with the development of the laser modulation scheme for quasi-simultaneous measurements of different quantities. In addition, necessary extensions to the spectrometer design are discussed. General concepts of an application-independent optical setup are described in sec. 3.3 and an error analysis is performed in sec. 3.4. The chapter concludes with a literature review and experimental measurements of necessary line coefficients, which significantly improves the quality of the data compared to databases. Parts of this chapter were published by the author in [E2].

3.1 Line Selection

The selection of suitable absorption lines highly depends on the thermochemical conditions and spatial restrictions of the combustion system to be investigated. The conditions are estimated to:

- temperatures in the range of 300 K – 2200 K,

- atmospheric pressure,

- concentrations of combustion relevant species in the range of:

Species	CO_2	H_2O	O_2	N_2	CH_4	CO	HCl	OH	NO	NO_2	C_2H_2
Min [ppm]	5 000	5 000	2 000	100	10	10	10	10	1	1	1
Max [%$_{Vol.}$]	90	20	30	80	100	15	0.5	0.1	0.1	0.1	0.05

These species are among the most relevant species within combustion research. Additional minor pollutants, in particular of solid-fuel combustion, such as sulfur dioxide (SO_2), sulfur trioxide (SO_3), potassium (K), nitrous oxide N_2O and mercury (Hg) as well as unburned hydrocarbon and hydrocarbon radicals, such as CH_3, CH_2 and CH are also often investigated and are accessible by absorption spectroscopy, but will not be considered in this work.

In principle, rotational-vibrational transitions with high absorption strength are preferable, as they promise lower detection limits, in particular for minor species. The strongest absorption lines can usually be found in the fundamental bands which are located in the mid-infrared region (MIR) from 3 μm to 8 μm wavelength. Despite current developments, in particular of quantum and interband cascade lasers, the laser and fiber technologies available in the mid-infrared are not as mature as technologies in the near-infrared region (NIR). Hence, transitions of the overtones in the near-infrared region from 0.7 – 3 μm will be used in this work. Plenty of literature is available on the selection of optimal absorption lines [104,115,162–166], depending on the accuracy of the line data and the availability of corresponding lasers. The most important criteria for a line selection are:

- high optical density over the whole temperature range,

- separation from absorption lines of the same or other species,

- low temperature dependence of the absorption lines for concentration measurements,

- highly different temperature dependence of two absorption lines for temperature measurements by two-line thermometry,

- availability of tunable lasers at the chosen wavelength,

- quality of the absorption line coefficients.

Practice has proven optimal results for a peak absorbance in the range of $\sim 10^{-3}-1$ OD_e (optical density based on natural logarithm). For a lower absorbance, the separation of the absorption from electrical noise at the absorption line root and laser background nonlinearities becomes complex. For absorbance levels significantly exceeding this range, electrical noise at the absorption line peak leads to severe systematical errors.

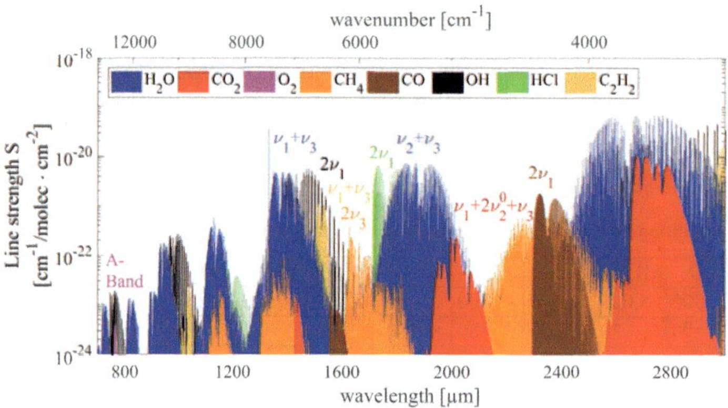

Fig. 3.1: Absorption line strengths of combustion-relevant species at 1000 K, simulated with HITRAN

Fig. 3.1 shows a simulated absorption spectrum of some of the aforementioned combustion-relevant species. The species concentrations were set to half of the maximally expected values. Methane poses an exception: as pure CH_4 is only relevant for non-premixed cases, a more representative concentration of 2.5 %$_{Vol.}$ was chosen. Nitrogen, as a symmetric molecule, does not possess a dipole moment and therefore does not have any rotational-vibrational transitions that can be excited in the NIR. NO and NO_2 do not have any significant absorption lines in the NIR, as the overtones are either in the MIR or too small. Therefore, these species will not be investigated in this work.

For each of the remaining combustion gas species, absorption lines were selected based on the aforementioned criteria and individual spectrometers were developed. In the following, the relevance of each species for combustion research will be discussed in brief and the results of the line selection will be presented. An approximate absorption path length L necessary for achieving peak absorbance levels in the aforementioned range will be given for each absorption spectrum. This information aids the development of the optical system (section 3.3).

Some of the developed spectrometers were successfully used by the author besides the investigations of this work for research in the fields of turbine inlet temperature measurements in pressurized gas-turbine combustion chambers [E3], fuel and water vapor mole fraction measurements in an IC-engine under motored [E4] and fired operation, tomographic 2D reconstruction of the thermochemical state in a diesel exhaust gas after-treatment system as well as for concentration measurements in pressurized nuclear disaster simulation rigs.

3.1.1 Water Vapor (H$_2$O)

Water vapor is one of the major combustion products through the combustion of hydrogen or hydrogen-containing fuels. In addition, it also occurs in the reactants due to humidity. H_2O has multiple overlapping hot- and cold-bands, which makes the species well suitable for universal temperature measurements using two-line thermometry [100,167]. H_2O concentration measurements can be used to determine the reaction progress of gaseous fuels and the progress of pyrolysis of solid fuels.

Due to the large number of normal modes, water absorption lines cover almost the entire NIR. The $\nu_1+\nu_3$ and $\nu_2+\nu_3$ combination-bands of H_2O around 1400 nm (7143 cm^{-1}) and 1850 nm (5405 cm^{-1}), respectively, have gained a lot of interest in combustion research [98,100,105,112,167–170]. The absorption lines in these regions are in general relatively strong, well separated from each other and absorption lines of foreign species and have widely different temperature dependences of the line strengths, due to a partial overlap with the $2\nu_1$ and $2\nu_2$ hot-bands (fig. 3.1), which feature significantly different lower-state energies. Furthermore, due to the proximity to the telecommunication S- and U-band, high-grade diode lasers and low-attenuating fibers are available in these regions. In particular, the $202\leftarrow303$ line at 7185.6 cm^{-1} in combination with the $660\leftarrow661$ line at 7181.16 cm^{-1} and the $212\leftarrow313$ at 5393.65 cm^{-1} together with the $982\leftarrow981$ line at 5394.2 cm^{-1} are very promising for temperature and water vapor measurements [105] [E3,E4].

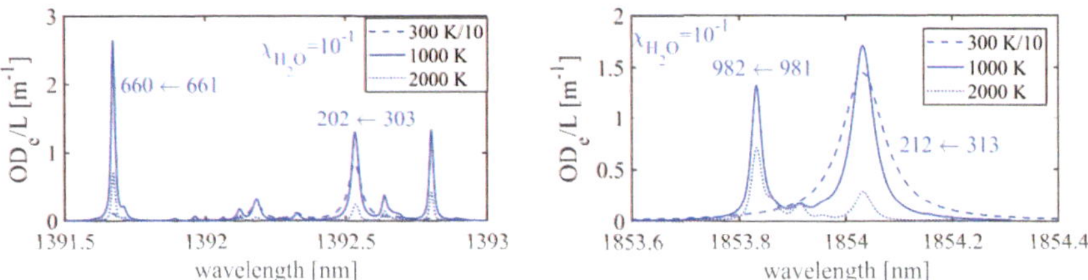

Fig. 3.2: Length-normalized absorption spectra of H_2O in the $\nu_1+\nu_3$ (left) and $\nu_2+\nu_3$ (right) combination bands. The spectra for 300 K temperature are scaled down for a better visualization.

Fig. 3.2 shows simulated spectra of these and adjacent absorption lines for exemplary combustion-relevant temperatures. In these and the following figures, spectra of interfering species are shown only where relevant and non-negligible. The $202\leftarrow303$ and $660\leftarrow661$ absorption lines have a fairly different temperature-dependence due to different ground state energies (140 cm^{-1} to 1045 cm^{-1}), which makes them well suitable for temperature and water vapor measurements. In the $\nu_2+\nu_3$ combination band, the ground state energies of the $212\leftarrow313$ and $982\leftarrow981$ lines differ even farther from each other (79 cm^{-1} to 2040 cm^{-1}), which promises even lower uncertainties in the temperature measurements. However, due to the high ground state energy, the $982\leftarrow981$ transition is barely separable from its partner line for temperatures below ~600 K. Hence, a temperature determination with the second line pair is only possible for temperatures above this threshold.

For an optimal absorbance with peak strength in the magnitude of $OD_e \sim 1$, an absorption length of the order of $L = 1$ m is required for both systems. For ambient temperatures, the simulated water vapor concentrations lead to peak absorbance significantly exceeding $OD_e = 1$. However, due to condensation, the water vapor concentrations in this region are also significantly lower than simulated, which makes the absorption lines suitable for measurements at ambient temperatures as well. Due to these considerations, the two absorption line pairs can be used for universal temperature measurements in the unburned gases, the reaction zone and the exhaust gases.

3.1.2 Carbon Dioxide (CO₂)

Carbon dioxide is one of the major combustion products. In oxy-fuel combustion, it is particularly relevant as it is recirculated into the unburned gases. Similar to water vapor, CO_2 allows to estimate the combustion progress of gaseous fuels. In addition, its concentration highly depends on the burnout rate of char as this substance mainly consist of carbon, which makes it interesting for measurements of the char burnout rate. Furthermore, carbon dioxide emissions are regulated through emissions trading. Carbon dioxide and water vapor dominate the heat exchange by radiation in gases. This is of high relevance throughout the entire combustion system, as it affects multiple processes, including the heating of unburned gases, the flame, the chemistry, the flow and the design of flue gas heat exchangers in power plants. Hence, knowledge of the species concentrations is crucial for modeling the radiative heat transfer rates in particular in oxy-fuel processes.

In the NIR, the 2.7 µm band of CO_2 offers the highest line strengths. However, this band is rarely used for measurements due to a poor availability of low-attenuation fibers and sensitive photodiodes in this region, which increases the complexity of the optical setup. In addition, the optical densities are too high for measurements in oxy-fuel processes. The $v_1+2v_2^0+v_3$ combination band around 2 µm (5000 cm^{-1}) on the other hand has already been used for CO_2 detection in combustion and gasification processes [98,100,100,102], as it lies within the range of the photodiodes and the line strengths are sufficient for oxy-fuel combustion processes. In this band, the R(18) line at 4991.26 cm^{-1} is of special interest, due to its line strength and separation from lines of other species (fig. 3.3). At elevated temperatures, adjacent high-temperature lines become apparent. The combination of the R(18) line with the R(39) line at 4990.66 cm^{-1} [171] is well suited for determination of the CO_2 temperature via two-line thermometry, due to their largely different ground state energies (133 cm^{-1} vs. 1276 cm^{-1}) and thus temperature dependences. A NIR two-line thermometry based on CO_2 has not been described in literature, as the concentrations are usually too low to lead to sufficiently high peak absorbance for low uncertainties. The significantly raised CO_2 concentrations of oxy-fuel combustion offer the unique possibility to evaluate the qualities of this technique through comparison with the H_2O-two-line thermometers. For a precise determination of the CO_2 absorption spectrum, interfering H_2O absorption lines have to be taken into account (fig. 3.3), which can be calculated from the water vapor concentration gained from the separate spectrometers.

For an optimal absorbance with peak strength in the magnitude of $OD_e \sim 1$, an absorption length around $L = 1$ m is required.

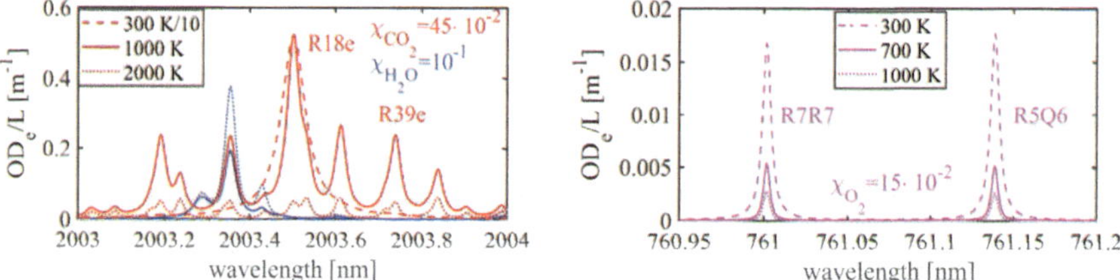

Fig. 3.3: Length-normalized absorption spectra of CO_2 in the v_1+v_3 (left) and O_2 in the A-band (right) combination bands for various temperatures. The spectrum of CO_2 for 300 K temperature is scaled down for a better visualization.

3.1.3 Oxygen (O_2)

Oxygen (more precisely dioxygen) is essential for combustion processes as it serves asoxidizer. In the inlet flows, the oxygen concentration is required to calculate the equivalence ratio, in the flue gas it defines the oxygen excess. In oxy-fuel combustion, zero oxygen excess is desired to avoid unnecessary waste of oxygen. Whereas local measurements of the O_2 concentration in the combustion region allow an identification of oxygen slip regions and the oxidizer to fuel ratio, global measurements allow to study the burnout rate of fuel.

Due to the symmetry of the molecule, O_2 cannot be excited in the NIR through fundamental vibrational-rotational transitions. However, a change in the rotational state of the molecule can be electronically excited through magnetic dipole-intercombination-transitions. The absorption line strengths induced by these processes are several orders of magnitude lower than for vibrational-rotational transitions. The A-band $(0\leftarrow 0)$ of the $^1\Sigma^+_g \leftarrow ^3\Sigma^-_g$ transition around 760 nm has been often used for O_2 measurements [99,172–174], as it offers the highest available line strengths in the NIR. In addition, the lines are well separated from each other and no other species have significant transitions in this region. High-quality tunable lasers and fibers are available for this band, as it is close to the visible region where diode lasers are used for illumination applications.

In particular, the R7R7 and R5Q6 transitions at 761.0 nm (13140.6 cm^{-1}) and 761.1 nm (13138.2 cm^{-1}) in the R-branch of the A-band are suitable for combustion research, as their line strengths have a comparatively low temperature dependence [175,176]. Nevertheless, the peak absorption of the transitions (fig. 3.3) is still significantly decreasing with temperature due to a reduced gas density and low ground state energy, which makes an O_2 measurement impractical for temperatures above 1000 K. Even for temperatures below this limit, an absorption length L of a few meters is required, which is challenging to establish in a combustion environment.

3.1.4 Methane (CH$_4$)

The reaction kinetics of the methane oxidation are comparatively well-known and the reaction mechanism necessary to describe the kinetics is among the most simple mechanisms [61,177]. Therefore, methane combustion is often chosen for generic setups. In oxy-gas combustion, methane is highly relevant as it is one of the main constituents of natural gas as fuel. The concentration of the gas in different locations of the flame or combustor allows to estimate the reaction progress. In oxy-coal combustion, a large share of the volatiles consists of methane. In both cases, methane residues in the flue gas allow to investigate the burnout rate. During combustion chamber ignitions, CH$_4$ concentration measurements are relevant for safety aspects due to deflagration events.

Methane has a large number of vibrational modes due to the high number of atoms in the molecule, which lead to CH$_4$ absorption lines throughout the entire NIR. The $2v_3$ stretching-vibrational overtone of CH$_4$ around 1650 nm [174] is well suitable, as it offers high absorption line strengths and is within the telecommunication U-band. Thus, high-grade lasers are available at low costs. In this band, the R(3) line-triplet at 1653.7 nm (6046.9 cm^{-1}) has already been used multiple times for methane detection [166,171,176,178], as it is free of interference with other species (fig. 3.4). For typical fuel gas compositions, an absorption length L of a few centimeter is sufficient, for a detection of flue gas residues, an absorption length around L=1 m is preferable.

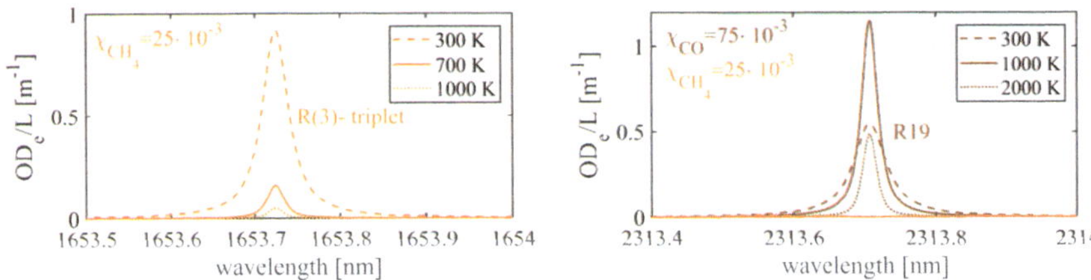

Fig. 3.4: Length-normalized absorption spectra of CH$_4$ in the $2v_3$ (left) and CO in the $2v_1$ (right) overtones for various temperatures.

3.1.5 Carbon Monoxide (CO)

Carbon monoxide is highly relevant for combustion research and the design of combustion systems. Emissions of this toxic pollutant are highly regulated and need to be avoided, as high concentrations require costly flue gas after-treatment. CO is primarily formed in fuel-rich environments, where an oxygen-depleting environment prevents an oxidation to CO_2. Besides during the combustion of char (subsection 2.1.4), high CO concentrations are also formed through an incomplete combustion progress. Some major reaction paths of the methane combustion, e.g., follow the intermediates of CH$_4$ → CH$_3$ → CH$_2$O → CHO → CO → CO$_2$ [177]. Herein, the reaction steps towards CO are comparatively fast, whereas the oxidation of CO + OH → CO$_2$ + H is slow. Hence, flame quenching throughout the reaction progress by an enthalpy loss, e.g. at cold superheater surfaces or cold airflows before reaching equilibrium conditions, primarily results in high CO concentrations. High CO concentrations near the superheater or heat exchanger surfaces can increase corrosion [179] as well as slagging [73] through a reducing atmosphere.

As described in subsection 2.1.4, CO plays a crucial role in the combustion of char, as the char is not reacting with molecular oxygen directly but through an CO-enriched atmosphere around the char particle [73]. In oxy-fuel combustion, carbon monoxide formation is particularly relevant, as the high concentrations of carbon dioxide intensify the reverse reaction of CO_2 → CO through the Boudouard-reaction ($CO_2 + C \rightleftharpoons 2CO$) or the water-gas shift reaction ($CO_2 + H_2 \rightleftharpoons CO + H_2O$). Measures have to be taken to reduce the CO concentrations affiliated with these reactions.

In principle, CO can be measured within the $3v_1$ stretching-vibrational overtone around 1600 nm, which lies within the telecommunication S-band. However, the absorption lines strengths are two orders of magnitude lower than within the first overtone around 2300 nm. This band has been used for various combustion-related investigations [99,100,102,111,115,171,174,178,180]. However, most CO absorption lines in this band overlap with non-negligible H_2O or CH_4 absorption lines. The R(19) transition at 4322.1 cm^{-1} (2313.7 nm) is a rare exception, as it is only interfered by one minor CH_4 line in the line flank, which can easily be accounted for within the post-processing. This absorption line also offers exceptionally high line strength and a low temperature dependency, which allows measurements throughout the entire range of combustion-relevant temperatures (fig. 3.4). For low CO concentrations, e.g. in the flue gas, an absorption path length around L=1 m is preferable. For high CO concentrations within the flame front, shorter absorption lengths in the order of a few centimeters are also feasible.

3.1.6 Hydroxyl Radical (OH)

Most reaction paths of hydrocarbons with molecular oxygen to CO_2 and H_2O rely heavily on OH as reaction partner or product [177]. Hence, without significant OH concentrations, a combustion does not take place. On the other hand, as OH has a relatively short lifetime (~ms) and quickly recombines outside its formation zone in the flame front, it is a formidable indicator of the flame front and heat release zone [23,33,178,181]. In addition, through its intense chemiluminescence it can easily be visualized [54,178]. In the post-flame region, a thermal equilibrium concentration of OH remains, which allows temperature measurements from quantitative knowledge of the OH concentration [154–156]. In char combustion, OH reactions play a less significant role, as carbon is oxidized with CO_2 to CO without the need of OH. However, the reverse oxidation of CO to CO_2 outside the CO atmosphere around the char particle relies on OH as a reaction partner [73]. In the flue gas analysis, OH concentrations allow for an identification of remaining combustion zones, as these zones temporally increase the equilibrium OH concentrations.

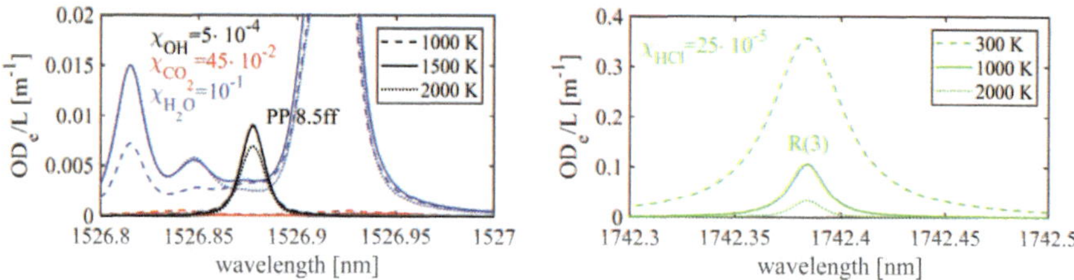

Fig. 3.5: Length-normalized absorption spectra of OH (left) and HCl (right) in the $2v_1$ overtones for various temperatures

OH has multiple overtones of the stretching vibration in the NIR. The first overtone ($2v_1$) around 1500 nm is the strongest, which makes it interesting for combustion research. The P 8.5ff absorption line at 1526.9 nm (6549.3 cm^{-1}) is comparatively free of interference by water vapor and is easily accessible through diode lasers in the telecommunication S-band (fig. 3.5) [178]. However, the absorption line strength is comparatively low, which causes a need for either long absorption lengths or an elaborate noise reduction through filtering or averaging of the signal. Measurements at an absorption length around L=1 m are possible, but longer absorption lengths are preferable.

3.1.7 Hydrogen Chloride (HCl)

Solid fuels, such as coal or biomass, contain high shares of volatile alkali compounds as well as chlorine [161,182]. In addition, some of the fuels contain embedded salts, e.g. KCl or NaCl. During combustion, these compounds are released into the gas phase and form gaseous hydrogen chloride (HCl), which can lead to disadvantageous effects on operation and performance of the combustion system through enhanced formation of deposits and high-temperature chlorine-induced corrosion in the superheater [183–185]. At the heat exchanger surfaces, alkali salts may form deposits via condensation or aerosol particles. Additionally, HCl is a toxic pollutant that forms hydrochloric acid in the presence of water. Hence, monitoring the HCl emissions in the flue gas of solid-fuel combustion processes is highly relevant. Local in-situ measurements provide insights into the formation of HCl in the devolatilization and the char burnout, which eventually leads to reduction concepts. In addition, as will be described in further detail in section 5.2, HCl is a formidable tracer in combustion processes due to its stability.

In principle, HCl has two relevant stable isotopologues, $H^{35}Cl$ and $H^{37}CL$, with a natural abundance ratio of 75.77 %$_{Vol.}$ to 24.23 %$_{Vol.}$. The mass difference of the two isotopologues induces a shift of the absorption lines in the same band by a few wavenumbers. Even though the bands of the two isotopologues overlap, the absorption lines can be separated from each other. Usually, $H^{35}CL$ concentrations are measured due to the higher abundance, and the overall concentration of HCl is calculated from the natural ratio. The $2v_1$ and $3v_1$ stretching-vibrational overtones of $H^{35}CL$ around 1750 nm and 1200 nm offer considerable transitions in the NIR, however, the $2v_1$-band is usually preferred due to its superior line strengths [101]. The R(3) absorption line at 1742.4 nm (5739.3 cm^{-1}) has already been used for combustion-related measurements by Orthwein et al. [106], as it is relatively strong even at adiabatic flame temperatures and separated from any other relevant absorption line (fig. 3.5). For concentrations in the magnitude of a few hundred ppm an absorption length of L=1 m is optimal, whereas for lower concentrations a few meters are preferable.

3.1.8 Acetylene (C₂H₂)

In the combustion of hydrocarbon fuels, the fuel is broken down into C_1-and C_2- hydrocarbon fragments prior to reactions to CO_2 and H_2O [186]. Fuel-rich conditions lead to acetylene formation in high concentrations (fig. 3.6 a) [187]. Acetylene reacts with CH to C_3H_3, which can form a first benzene ring (fig. 3.7). These rings can grow polycyclic aromatic hydrocarbon- (PAH) molecules, which can form particle-like structures. These structures grow through conglomeration and surface growth by addition of acetylene, which eventually leads to the formation of soot, i.e. macromolecule particles [188]. Hence, acetylene is an important precursor for soot inception and soot growth in fuel-rich conditions. As soot is possibly carcinogenic, emissions are regulated and a formation is to be avoided. Here, in-situ measurements allows insights into an improvement in the combustor design. In solid fuel combustion, acetylene is one of the major constituents of the volatiles and allows to identify the onset of pyrolysis.

C_2H_2 has rarely been measured in the NIR, as the line strengths are significantly lower than in the mid-infrared fundamental bands. However, the reduced complexity of the experimental apparatus make the v_1+v_3 combination band around 1540 nm a more practical region [97]. As this combination band lies in the telecommunication C-band, high-quality lasers are available. The P(23) line at 6495.91 cm^{-1} is sufficiently separated from H_2O hot-bands (fig. 3.6b) and has a comparatively low temperature dependence. However, the absorbance significantly drops above 1000 K, which impedes in-flame measurements. In flue gases, however, measurements are feasible for an absorption length around L=1 m.

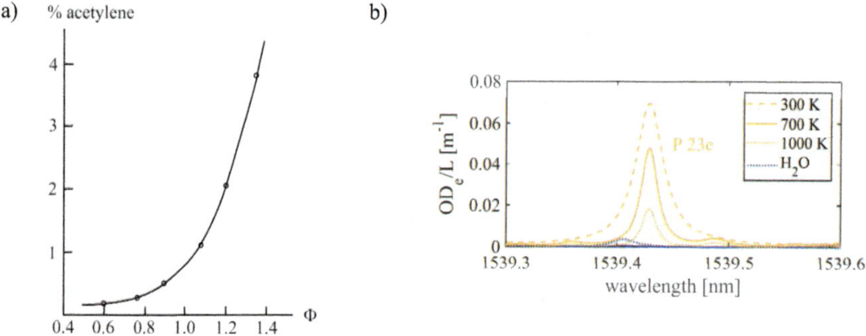

Fig. 3.6: Acetylene formation in CH$_4$-O$_2$ flames as a function of equivalence ratio [187] (a); length-normalized absorption spectrum of C$_2$H$_2$ in the $\nu_1+\nu_3$ combination band for various temperatures (b).

Fig. 3.7: Steps of PAH formation [188]

3.2 Data Acquisition System

This section describes the setup of all measurement system components with the exception of the optical elements, which will be described in the section hereinafter. The overall measurement system for one species will be referred to as a spectrometer. First, an overview of the selected lasers is given, followed by descriptions of the laser modulation scheme and the hardware. The section is finalized by an introduction into the post-processing procedure of spectral fitting.

3.2.1 Laser Selection

The absorption spectra are measured by scanning the wavelength of a narrowband diode laser across the corresponding spectral range. The central wavelength of each diode laser therefore needs to be tunable to the absorption line central wavelength via temperature tuning. Practice in spectrometer design has proven additional requirements for the lasers:

- a side-mode suppression ratio of at least 25 dB to eliminate effects of side-band absorption,

- a laser line width 20 MHz or less to avoid a convolution of the laser spectral profile with the absorption line profile, which has a line width in the order of GHz,

- a sufficiently wide tunability through injection current to cover the desired spectral range, at least multiple times wider than the absorption line width in order to be able to identify the background

- tunability through the injection current with a bandwidth $>100\,\text{MHz}$ to allow a high measurement repetition rate ($\sim$kHz),

- sufficiently high output power ($>100\,\mu$W),

- a correlation between the injection current and the wavelength as well as the output power that can be described by simple modeling functions,

- high stability and long-term reproducibility of the tuning over several years for achieving a calibration-free measurement system.

These requirements are fulfilled by the laser types described in section 2.2.5. Hence, the measurement system utilizes a set of vertical cavity surface emitting diode lasers (VCSEL), distribute feedback lasers (DFB) and novel discrete mode lasers (DM), which are listed in table 3.1 together with the corresponding absorption lines and measured quantities. With the exception of the O_2-measurement system, all lasers are fiber coupled for robust measurements in the harsh environment of the combustion systems and a simplified operation of the system.

The spectral separation of the two O_2-absorption lines and the H_2O-absorption lines for two-line thermometry at 1392 nm require the wide tunability of VCSELs. The two lines of the 1854 nm thermometer are spectrally closer, hence a DM with a lower tunability but higher output power is sufficient.

Table 3.1: Summary of the lasers used within the spectrometers and corresponding transitions under investigation

System	#	laser (central wavelength/ type/ manufacturer)	target species/ main transitions/transition wavelength [nm] (wavenumber [cm^{-1}])	measured quantities
I	1	2314 nm/ DFB/ nanoplus	CO/ R(19)/ 2313.7 (4322.1)	c_{CO}
	2	2004 nm/ DFB/ nanoplus	CO_2 / R39e/ 2003.7 (4990.7)	c_{CO2}
			CO_2 / R18e/ 2003.5 (4991.3)	T_{CO2}
	3	1854 nm/ DM/ Eblana Photonics	H_2O/ 2 1 2 $\leftarrow$ 3 1 3/ 1854.0 (5393.7)	c_{H2O}
			H_2O/ 9 8 2 $\leftarrow$ 9 8 1/ 1853.8 (5394.2)	T_{H2O}
II	4	1742 nm/ DM/ Eblana Photonics	H^{35}Cl/ R3 1742.4 (5739.3)	c_{HCl}
	5	1653 nm/ DFB/ NEL	CH_4/ R(3)- triplet/ 1653.7 (6046.9)	c_{CH4}
	6	1539 nm/ DFB/ NEL	C_2H_2/ P 23e/ 1539.4 (6495.9)	c_{C2H2}
	7	1527 nm/ DFB/ NEL	OH/ PP 8.5ff/ 1526.9 (6549.3)	c_{OH}
	8	1392 nm/ VCSEL/ Vertilas	H_2O/ 2 0 2 $\leftarrow$ 3 0 3/ 1392.5 (7181.2)	T_{H2O}
			H_2O/ 6 6 0 $\leftarrow$ 6 6 1/ 1391.7 (7185.6)	c_{H2O}
III	9	761 nm/ VCSEL/ Philips	O_2/ R5Q6/ 761.1 (13138.2)	c_{O2}
			O_2/ R7R7/ 761.0 (13140.6)	

The lasers are categorized into systems depending on their central wavelength. Lasers above 1800 nm (System I) require fibers with a core diameter of $11\,\mu$m and a Ge-doped silica core for single mode laser transmission at low attenuation. These fibers (Thorlabs SM-2000), however, show high attenuation for wavelengths below 1800 nm. For lasers with a shorter wavelength (System II), pure silica fibers with $8.2\,\mu$m core diameter (Thorlabs, SMF-28) are employed, which cannot be utilized above 1800 nm due to a multi-modal transmission in this spectral range. Below 1300 nm, these fibers likewise show high attenuation. Here, a free beam laser is used (System III).

3.2.2 Laser Modulation and Multiplexing

In turbulent combustion systems, simultaneous or quasi-simultaneous measurements at the same location are required for a determination of the species concentration using the temperature measured with a different spectrometer. Quasi-simultaneous measurements also allow to calculate correlations between different species. Furthermore, they significantly reduce the measurement effort during measurement campaigns, as the thermochemical state can be probed all at once instead of sequentially for each species. In principle, two techniques allow quasi-simultaneous measurements along the exact same laser path:

1. Spectral multiplexing: All lasers operate simultaneously. Before entering the sample volume (i.e. the combustor or flame), the lasers are combined into one beam using filters, dispersive elements or fiber-coupled wavelength division multiplexer (MUX) [189]. After crossing the sample volume, the different lasers are spectrally separated using the elements inversely and focused onto individual photodiodes. While this technique allows simultaneous measurements at a high repetition rate in combustion environments [100,102], the optical elements necessary for a wavelength multiplexing and de-multiplexing (DEMUX) are expensive and need to be specifically designed depending on the lasers to be used. For a large number of lasers, this becomes challenging as the elements pose restrictions on the laser wavelengths, such as the spectral range or separation of different lasers.

2. Temporal multiplexing: In this technique, the lasers are operated one after another. Similar to spectral multiplexing, the lasers are combined into one beam before entering the sample volume. However, after crossing the volume, all lasers are focused onto one photodiode and do not need to be spectrally separated, as only one laser is operated at a time (fig. 3.8) [190,191]. This reduces the maximally possible measurement repetition and spectral separation of the lasers due to the detector sensitivity, but also the complexity, as no optical elements are required for wavelength separation. For sufficiently high repetition rates, this technique allows quasi-simultaneous measurements.

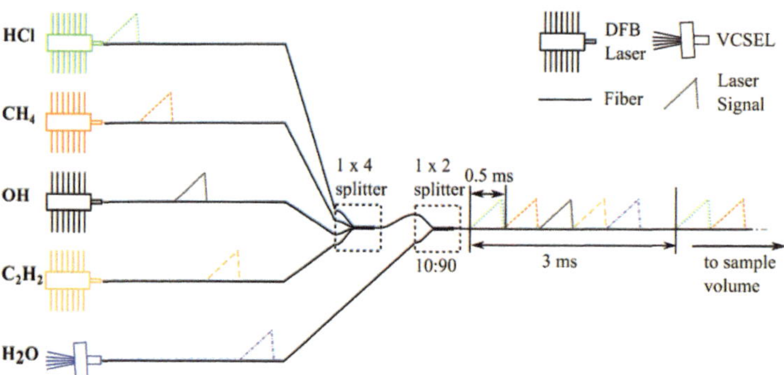

Fig. 3.8: Temporal multiplexing scheme of System II

Temporal multiplexing has been chosen in this work, as the measurement system is desired to be flexible with respect to extensions by additional spectrometers. Fig. 3.8 exemplarily shows the modulation scheme for System II: sequentially, each laser is tuned over its spectral range and emits light, while the injection current of the other lasers is below the emission threshold. During emission, the injection currents of all lasers are modulated in a 90 % asymmetric triangle shape, with spectra being recorded during the up-ramp. This allows a high spectral resolution, but saves the lasers from rapid current changes during the down-ramp. Each ramp has a duration of approximately 0.5 ms. The cycle repeats after 3 ms, which allows to operate up to 6 lasers quasi-simultaneously. Hence, the repetition rate of the measurement system and upper limit of the measurement base frequency is 330 Hz. The laser fibers are combined into one single fiber (section 3.3.1) which guides all lasers towards the sample volume.

Table 3.2: Controller setup and photodiodes

System	#	Laser	Function generator	Current Controller	Temp. controller	PD
I	1	2314 nm CO	Tektronix AFG 3022B	SRS LDC 500		Hamamatsu G12183
	2	2004 nm CO_2	LeCroy Arbstudio 1104	LDC 8002 Thorlabs		
	3	1854 nm H_2O				
II	4	1742 nm HCl			Thorlabs TED 8020	Hamamatsu G12181
	5	1653 nm CH_4	Tektronix AFG 3022B			
	6	1527 nm C_2H_2	LeCroy Arbstudio 1104			
	7	1529 nm OH				
	8	1392 nm H_2O		Thorlabs LDC 200C		
III	9	761 nm O_2	Tektronix AFG 3022B	SRS LDC 500		Thorlabs FDS010

Each laser is temperature- and injection current-controlled. The injection current controllers are modulated by function generators, which are synchronized using a pulse generator (GW Insteck SFG-2004). The setup of the controllers and function generators is illustrated in fig. 3.9 and the corresponding models listed in table 3.2.

System I is operated similar to System II but with 3 lasers only, leaving space for further extension into the MIR. System III, which only consists of one laser, is not multiplexed. However, in order to be comparable to the other systems, it is also tuned during a 0.5 ms period in a repetitive cycle with 330 Hz base frequency. The residence time distribution measurements (section 5.2) pose an exception: Here, the system does not need to operate quasi-simultaneously, as only one species is investigated at a time. Therefore, the laser is operated at 2 kHz base frequency without multiplexing to increase the measurement rate.

3.2.3 Data Acquisition Hardware

After propagation through the sample volume, i.e. the flame or combustor, the laser light is focused onto a photodiode. Each measurement system utilizes a different photodiode, which are listed in table 3.2, to achieve a high wavelength-dependent sensitivity. Each photodiode signal is amplified (Femto DHPCA-100 transimpedance amplifier) and recorded using a National Instruments PXIe-6124 A/D converter, which is triggered by the pulse generator. The phase-locked acquisition of the photodetector signals is performed via an in-house LabVIEW software. During all measurements, the sample frequency of the photodiode signal was 4 MS/s. During some measurements, it was necessary to record multiple laser beams in parallel.

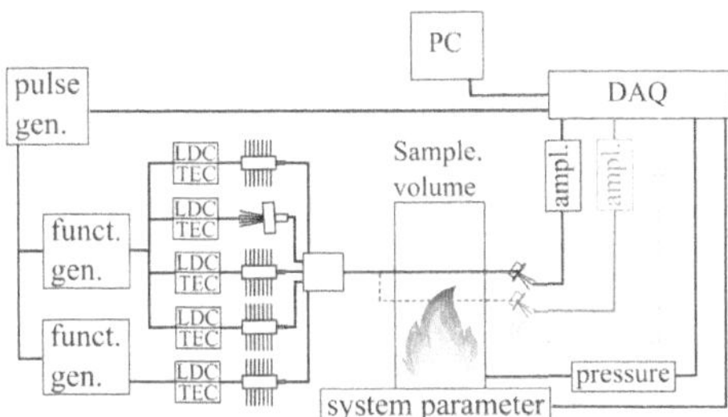

Fig. 3.9: Elementary measurement setup

Together with the amplified photodiode signals, system parameters of the facility and (in case of RTD-measurements) the valve status were recorded. Depending on the application, these include laser and burner position, system temperatures and valve triggers. The latter were recorded using the NI PXIe-6124 card, whereas the first parameters were acquired together with the pressure transducer signals using a slower NI-9215 A/D converter at 1 kS/s. The pressure in the sample volume was measured using different application-dependent pressure transducers:

- Line data measurements:

    - <1 kPa: CMR 363 pressure sensor (Pfeiffer Vacuum); measurement uncertainty 0.2% of the measured value,

    - 1–35 kPa: UNIK 5000 (GE); measurement uncertainty 0.07 kPa,

    - 35–1034 kPa: Model 8262-5150 (Burster); measurement uncertainty 0.5 kPa.

- Wolfhard-Parker Burner and Oxy-Fuel Burner: 60–110 kPa; PTB110 (Vaisala), measurement uncertainty 0.03 kPa,

- WSA-Technikum: 35–1034 kPa: Model 8262-5150 (Burster); measurement uncertainty 0.5 kPa.

A high-performance desktop-PC initializes the data acquisition and saves the data.

3.2.4 Post-Processing and Spectral Fitting

Prior to measurements at the different test rigs, each laser is characterized by placing a sapphire etalon into the laser beam. Partial reflections at the plano-parallel etalon surfaces cause multiple-beam interference, so called *fringes*, which impress a sinusoidal alteration onto the laser signal. The shape of this function is dependent on etalon-characteristics, which allows a transformation of the measured laser signal from the temporal space to the spectral space via the tuning function $\partial v/\partial t$ (subsection 2.2.3).

Voigt-shaped multiline-models of the absorption lines in combination with a polynomial laser baseline model are fitted to each scan of the experimental data using an in-house software based on LabVIEW [178]. A Levenberg-Marquardt algorithm ensures a quick convergence by regression of the least squares between the measured data and the model. Several Voigt functions are fitted simultaneously, which allows to not only evaluate the target absorption line but also to correct for the absorption background by absorption lines of the same or other species. The background polynomial accounts for background emission and transmissivity fluctuations. Plenty of literature is available on the fitting procedure [174,192,193], hence, only the most important steps are described here. Where necessary, individual changes of the procedure, such as the number of spectra used for averaging, are denoted in the each chapter.

The Doppler broadening of each Voigt-profile is calculated from the measured temperature and HITRAN/HITEMP spectral data. Despite the collisional broadening coefficients determined in section 3.5, the temperature coefficient of the main absorption lines as well as most broadening coefficients of the background absorption lines are poorly known. Hence, the collisional broadening must fitted for the complex oxy-fuel gas matrix, which significantly differs from air. For measurements with a line peak absorbance below 10^{-2} OD$_e$ (e.g. OH and C_2H_2 measurements), the optical density is of similar amplitude as the baseline noise. To partially compensate for this effect by noise reduction through smoothing, the bandwidth of these signals is reduced via a Savitzky-Golay filter of 7th polynomial order and a 14 pixel window size (450 pixel length of spectral window), which filters white noise [194–196].

Various of the measured spectra, in particular at high temperatures, consist of several hundred absorption lines. For example, the combination of the HITRAN2016 [88] and HITEMP2010 [89] databases list > 20 000 H_2O absorption lines for the 1392 nm spectrum alone, most of which are several orders of magnitude too small to play a significant role. To reduce the computational effort in the post-processing, about 120 of these are identified to be relevant in the temperature range from 300 K to 2500 K. Here, relevant is defined as having a relative line strength which exceeds a ten thousandth of the strongest transition in this spectral range for any temperature. The same procedure is performed for the 1854 nm H_2O and the 2004 nm CO_2 spectra.

For the temperature determination by two-line thermometry, the absorption lines are clustered depending on their lower state energy. One set of low and one set of high temperature lines are fitted simultaneously onto the spectrum. The absorption areas and collisional broadenings of each of the sets are optimization coefficients which are coupled within one set but independently fitted from the other set. This procedure requires prior knowledge of the temperature, as the shape of the spectrum within one set is temperature-dependent. The temperature is gained via a recursive algorithm: the temperature from the previous measurement is utilized as an initial value. After fitting, the new temperature, calculated according to eqn. 43, serves as an updated value. This iterative algorithm ensures a fast and stable convergence of the temperature. Measurements at a reference cell have shown, that three iterations of the algorithm are sufficient for the temperature to deviate < 1 K from the finite value, even for temperature changes of several hundred Kelvin.

3.3 Design of the Optical Setups

A fundamental component of the spectrometers is the optical setup, which combines the laser light into one beam and guides it into and out of the sample volume. For a reliable operation in the combustion chamber, this setup needs to fulfill several requirements:

- fiber-coupled combination of the lasers for high stability in hostile environments,

- insensitivity to beam-steering caused by temperature gradients in the combustor,

- elimination of parasitic absorption by species in the optical setup,

- simple beam alignment for a quick assembly,

- scalable design for an operation at both laboratory generic flames and close-to application combustors. In particular, large-scale combustion systems pose strict limitations to the optical access, which needs to be accounted for.

In this section, the basic optical setup will be explained. However, this setup requires application-dependent modifications, which are described in the respective chapters.

3.3.1 Design of the Fiber Setups

Wavelength-dependent multiplexers [189] offer a comparatively laser-loss free possibility to combine several lasers, but struggle with low flexibility with respect to the chosen wavelengths of the lasers. Optical fiber-splitters on the other hand, if reversely used, are largely wavelength-independent. In these splitters, the light intensity of one input is equally split into multiple output fibers, most commonly by fiber splicing (fused biconic tapering) or fiber-coupled Bragg-gratings (planar lightwave circuits). Splitters of both techniques can be reversely operated with the drawback of significant laser intensity loss. Besides the intensity loss of the optical element itself, which is typically in the order of a few percent, the reverse operation only combines a fraction of $1/n_{input\ fibers}$ into the output fiber. Hence, a reversely operated equally coupling 1x4 splitter allows less than 25 % of each laser to pass.

In this work, this characteristic of wideband optical fiber splitters is used for a combination of the lasers and a simultaneous intensity matching of the laser power. The different laser types significantly differ in output power, ranging from ~100 μW (VCSEL) to ~10 mW (DFB and DM). For an optimal utilization of the amplifier's dynamic range, the difference in power of each laser should be less than one order of magnitude (the photodiode sensitivity is roughly wavelength-independent). Hence, the intensity of the DFB and DM needs to be significantly reduced. Fig. 3.8 exemplarily shows the utilized splitter setup for System II: The DFBs are equally coupled into one fiber by using a 1x4 splitter (Thorlabs, TWQ1550HA). Afterwards, they are combined with the VCSEL using a 1x2 splitter, which splits the intensity by a 90:10 ratio (Thorlabs, TW1550R2A2). Thus, the intensity of the DFB is reduced by >97.5 %, whereas >90 % of the VCSEL intensity is allowed to pass. Fig. 3.10 shows the photodiode signals of the System II lasers after amplification, where, due to the intensity matching, the voltages are in the same order of magnitude. The HCl laser is weaker, as its wavelength has a higher attenuation in the splitter.

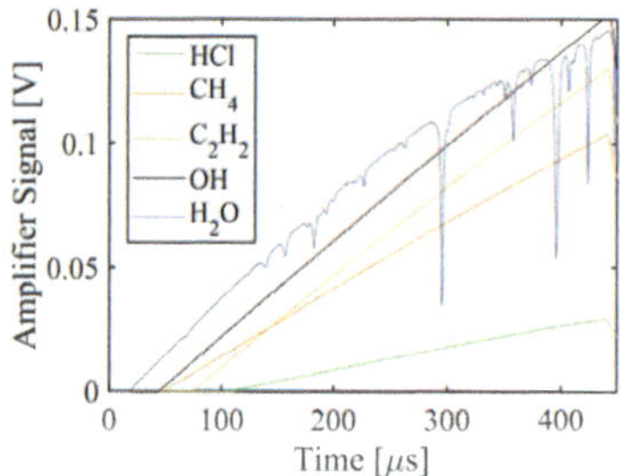

Fig. 3.10: Intensity-matched and combined laser signals of System II after absorption within the Oxy-Fuel Burner. The sequential laser signals are merged for a better visualization.

The lasers of System I are combined similarly. The 1854 nm H_2O and the CO_2 laser are combined by using a SM-2000 1x2 splitter (50:50, Thorlabs, TW2000R5A2A). The output fiber is connected to a second splitter of the same type, but 75:25 splitting ratio (Thorlabs, TW2000R3A2A), which is also connected to the CO laser. Hence, 25 % of this laser and 37 % of the first lasers are preserved. After combining the laser signals, the lasers are collimated and guided through the sample volume.

3.3.2 Design of the Absorption Path

The absorption path can be designed to have the laser collimation system and the photodiode at the same or at opposing sides of the sample volume. These designs are referred to as single-sided or double-sided setup. Whereas the first design allows an even number of crossings through the sample volume, the latter uses odd numbers. This work utilizes single-sided double- and multi-pass as well as double-sided single-pass (fig. 3.11).

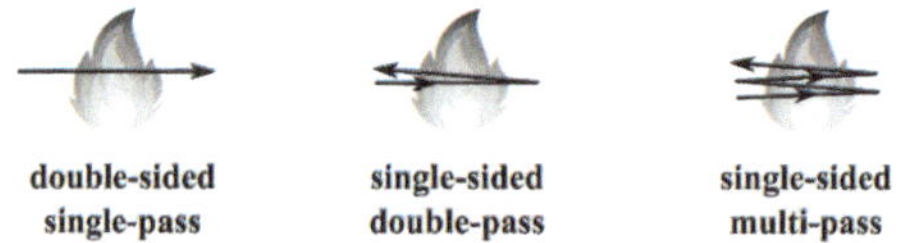

Fig. 3.11: Basic absorption path designs

The single-pass setup is utilized where a simple design with independent adjustment of the collimator and photodiode is needed. However, it comes with the drawback of a short absorption path-length. It is used during experiments at the Wolfhard-Parker Burner (chapter 4) due to its high spatial resolution. The single-sided setups require less optical access, as the large optical elements can be mounted on one optical port. It is used in applications with lower absorbance, as it offers a longer pathlength. However, the spatial resolution is lower, as the different reflections need a minor tilt or offset to be separable from the initial beam. Due to its robustness, it is used at the Oxy-Fuel Burner (chapters 5 & 6) and the Technikum (chapter 7). The multipass setup is utilized for applications with very low absorbance, but is less robust to beam-steering and misalignments. Particles within the sample volume also create an enhanced attenuation in this setup. Hence, it is mainly used for experiments at the Oxy-Fuel Burner under Oxy-Gas operation (chapter 6).

For single-sided setups, the combination of a retroreflector and an off-axis parabolic mirror (OAP) has proven its distinguished robustness [197]. Independently of the angle of incidence, the retroreflector reflects the laser into the same direction it came from (fig. 3.12). In contrast to a concave mirror, it does not change the collimation of the laser and the tilt does not need to be adjusted. The retroreflector induces a small offset between the incident and reflected laser beams. Taking the same path twice reduces effects of beam-steering. An OAP focuses collimated light onto the same position, independently of the laser position. Collimated laser light can be guided towards the reflector from a hole in the central OAP axis. Using this combination, the two rotational angles of the OAP only need to be adjusted to guide the laser to the retroreflector and the photodiode to be placed within the focus of the OAP, which requires low adjustment effort. In addition, the combination is resistant against thermal expansions causing misalignment of optical components.

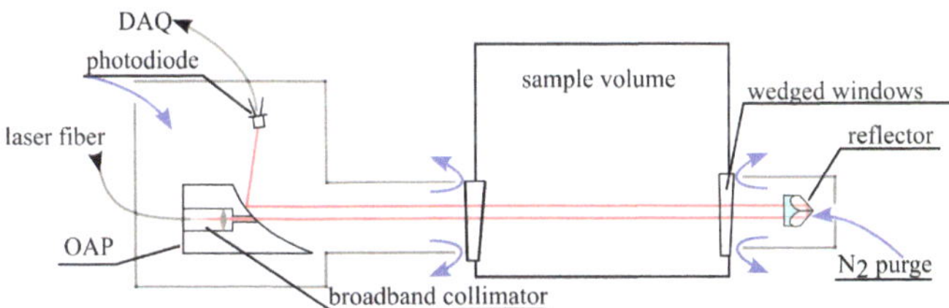

Fig. 3.12: Basic design of the single-sided doublepass setup. Planar mirrors at the sides of the sample volume, if necessary, can extend the path length.

If necessary, the laser path can be extended to a multi-pass setup by using a set of planar mirrors on each side of the sample volume. While this setup is less stable than multi-pass cells, such as White- or Herriot-cells [198,199], it can easily be applied to many combustion chambers at low component costs.

Miniaturized versions (0.5″ optical components) of the single-sided doublepass and the singlepass setup have been used for IC-engine in-cylinder [E4] and gas-turbine combustion chamber [E3] measurements by the author.

To achieve high accuracies, the absorption path-length in the sample volume needs to be known precisely. Hence, the optical ports into and out of the sample volume are covered by windows. Here, the retroreflector has an additional advantage, as it does not necessarily need to be covered by a window to create a well-defined absorption path onset. Independently of the design of the optical path, all components are entirely housed and purged with a high-purity N_2-flow to eliminate parasitic absorption of H_2O, O_2 or CO_2 in the free beam outside of the sample volume. The same flow is used to cool the probe or combustor windows, where necessary. During the measurements, fringes between optical elements (e.g. mirrors, lenses, window surfaces) need to be strictly avoided, as they generate sinusoidal patterns on the absorption spectra which make a separation from absorption lines more complex. Hence, all windows are wedged or anti-reflective (AR) coated to guide reflections out of the laser path.

3.4 General Error Analysis

The main quantities measured within this work are the gas-phase temperature and the species concentration. The measurement uncertainties of these quantities are calculated from the Gaussian error propagation of eqn. 41 and eqn. 43. In principle, the different contributions to the error can be separated into systematic errors affecting the accuracy and random errors, which reduce the precision. However, some effects, such as absorption profile fitting deviations and pressure measurements, contribute to both the random and systematical error. As a separation of both effects is complex, a combined measurement uncertainty will be calculated in this work.

The following parameters contribute to the temperature measurement uncertainty based on two-line thermometry:

- Absorption line area A: The uncertainty of this quantity is estimated from the signal-to noise ratio (SNR), which is derived from the ratio between the absorption line peak optical density and the standard deviation of the residual between the measured data and the fit.

- Absorption line strength S: The uncertainty is taken from the databases or the literature, depending on the lowest available uncertainties.

- Lower-state energy E_i: The databases give no information on the uncertainty of this quantity. However, from the low uncertainty of the absorption line positions which is also linked to the lower-state energy, it can be inferred that the error of this quantity can be neglected.

The species concentration determination is additionally influenced by uncertainties in:

- Temperature T: The uncertainty is estimated by the uncertainty in two-line thermometry.

- Pressure p: The uncertainty is taken according to the pressure transducer specifications.

- Absorption length L: This parameter is estimated to 1 mm for all applications with the exception of the multi-pass setup at the Oxy-Flame Burner, where it is increased to 4 mm due to the higher number of beam circulations. At the Wolfhard-Parker Burner, this induces a larger relative uncertainty which is due to the lack of windows limiting the path length.

- Laser tuning $\partial v/\partial t$: With the exception of the line parameter measurements, this factor is neglected, as it has a significantly lower contribution compared to other influences. For the line parameter measurements, the uncertainty is estimated from the uncertainty in the Etalon length.

As the errors of some of these quantities depend on many application-relevant parameters, such as electric noise levels in the environment, achieved quality of the measured spectra, species concentration and gas temperature uncertainties, a quantitative error analysis is performed for each application separately in the respective chapter. The errors in line coefficient measurements, which have different constituents than the thermochemical quantities, are discussed later in this chapter.

In time-dependent applications, such as turbulent systems, the state of the system is constantly fluctuating with a standard deviation around the mean, independent of the measurement system. Throughout the turbulent applications of this work, the standard deviations of the measurements due to process fluctuations, $\sigma_{\chi,exp}$ and $\sigma_{T,exp}$, were usually significantly larger than the measurements uncertainties, $\sigma_{\chi,fit}$ and $\sigma_{T,fit}$. Hence, unless noted otherwise, the error bars in turbulent systems indicate the process fluctuations. In systems with low process fluctuations compared to the measurement uncertainties (e.g. the laminar Wolfhard Parker Buner), both quantities are reported.

3.4.1 Inhomogeneous absorption paths

One fundamental requirement for TDLAS is a homogeneous temperature distribution along the absorption path. If this demand is not met, the path-integrated temperature measured via two-line thermometry will be somewhere in between the minimal and maximal temperature along the absorption path, non-linearly depending on the absorber and temperature distribution. Additionally, it will also falsify the concentration measurement. Tomographic reconstructions based on measurements from multiple angles [103,170,200,201] or hyperspectral profile fitting [50,109,110,169], based on measurements of multiple lines of multiple species, are promising techniques to measure in inhomogeneous sample volumes. However, these techniques exceed the scope of this work.

Hence, absorption paths are chosen along homogeneous thermochemical states. At the Wolfhard-Parker Burner, the laser is guided along a symmetry axis, and at the Oxy-Fuel Burner through the flue gas, where measurements and simulations[25] show a well-mixed state. Measurements at the WSA-Technikum were also mainly performed in the well-mixed flue gas downstream the burner, but additionally at upstream locations with increased systematic uncertainties.

3.5 Fundamental Line Coefficients

For low measurement uncertainties, the availability of high quality line data is crucial. In this section, the line coefficients of the chosen absorption line available in the HITRAN and HITEMP database are analyzed. For coefficients with unacceptably high uncertainties, literature is reviewed for improvements (first subsection) or, where necessary, the uncertainty of the parameters is significantly reduced through experimental investigations within two measurement cells (second and third subsection).

The HITRAN/HITEMP databases [88,89,202] list a set of six coefficients for each absorption line, which describe the absorption spectra (section 2.2):

1. *Absorption line strength S*: this coefficient is the intensity of the temperature dependent line strength $S(T)$ at reference temperature (296 K) and allows to calculate $S(T)$ through eqn. 31. It is the most important quantity for accurate measurements of the thermochemical state and will be analyzed in detail.

2. *Absorption line central position v*: as TDLAS does not rely on absolute line positions, this coefficient has a low relevance and will not be reviewed. Additionally, the HITRAN/HITEMP data quality is usually sufficiently high for most applications.

3. *Self-induced collisional broadening γ_s*: even though the overall collisional broadening can be fitted, the ability to calculate the width is important for a stable and accurate modelling of the overall collisional broadening (eqn. 35). In particular at low absorbance, noise or inaccurate background corrections significantly falsify a fitted collisional broadening and thus the absorption line area, which causes large errors in the measured species concentration or temperature. Hence, this parameter is relevant and will be reviewed.

4. *Air-induced foreign collisional broadening γ_{Air}*: In former HITRAN versions until HITRAN2012, only the broadening coefficients by air are listed. However, for a comprehensive absorption line description, in particular in oxy-fuel combustion where the gas composition significantly differs from air, the foreign broadening of all relevant species has to be considered. Therefore, HITRAN2016 starts to list broadenings by CO_2, He and H for a few selected molecules. While the latter two are irrelevant for oxy-fuel combustion, the first one is only

---

available for the C_2H_2 line. Hence, a comprehensive review of all relevant collisional broadening coefficients is unavoidable.

5. *Temperature-induced collisional broadening of air n_{Air}:* For accurate modeling of the absorption line shape at high temperatures, the temperature broadening coefficients from all species are relevant, too. However, as the measurements of these coefficients are complex due to the high necessary temperatures, they exceed the scope of this work and will only be discussed in brief.

6. *Air-induced pressure shift δ_{Air}:* Similar to the collisional broadening coefficients, pressure shifts are particularly relevant at low absorbance, where, using a *line-lock*, knowledge of the shift parameters allows a more precise modeling of the absorption line. Additionally, pressure shifts are relevant for complex spectra with multiple lines that drift independently of each other. The shift caused by all relevant species instead of 'air only' will be reviewed, due to their importance for oxy-fuel combustion.

Additional line-shape factors, such as Dicke narrowing coefficients or collisional-shift induced Fermi-resonances, are insignificant in combustion processes at ambient pressure and will be ignored. The uncertainties of the lower state energy E_i, as described in the previous section, are also not analyzed.

3.5.1 State of Research

The literature on each of the coefficients will be reviewed in the following, as the HITRAN/HITEMP databases often offer unsatisfactory uncertainties.

Absorption line strength S

The line intensity uncertainties of the chosen absorption lines in the HITRAN2016 database range from <1 % to >20 %, with the error of most absorption lines in the range of 1–2 %. Errors below 1 % are desired, whereas errors exceeding 2 % are considered unacceptably high and need to be improved where possible. Table 3.3 lists the coefficients with the lowest uncertainty found in literature. For the high temperature lines of H_2O at 1843 nm and CO_2, no literature values where found, as they can only be measured at experimentally challenging temperatures. The same holds for OH, which is additionally quickly recombining. For these as well as the CO and O_2 transitions, which are already known with low uncertainty, the HITRAN2016 coefficients were used. For all other transitions, literature values were found and used within this work, as they significantly reduced the uncertainty. With the exception of OH, all coefficients are now known with an uncertainty < 2 % with the majority even below 1 %. For an experimental improvement of the OH line data, a more complex apparatus is required [203], which exceeds the scope of this work.

While the spectroscopic HITRAN/HITEMP databases provide relatively accurate data on position and line strength of major absorption lines, the pressure shift of the absorption line and the pressure broadening coefficients show large uncertainties. The databases are intended for environmental applications and therefore mostly provide information on the broadening and shift caused by air as a foreign gas only. This is insufficient for oxy-fuel combustion, as the combustion atmosphere has a significantly different composition. The collisional broadening as well as pressure shift coefficients of each transition of a molecule depend on every species in the molecule's ambience. This leads to a large number of necessary parameters. Here, only the most important contributions in literature will be reviewed.

Table 3.3: Line strengths and uncertainties of the selected absorption lines listed in HITRAN2016 and comparison to the best available literature

System	#	target species/ transition	HITRAN2016 [26]	HITRAN2016 error [%]	Literature [26]	Literature Error (%)	Technique/ Reference
I	1	CO/ R(19)	$4.554 \cdot 10^{-22}$	<1			
	2	CO_2/ R39e	$9.405 \cdot 10^{-24}$	1–2	na[27]		
		CO_2/ R18e	$1.292 \cdot 10^{-21}$	1–2	$1.278 \cdot 10^{-21}$	0.1	TDLAS [204]
	3	H_2O/ 212 ← 313	$2.587 \cdot 10^{-20}$	1–2	$2.581 \cdot 10^{-20}$	< 1	FTIR [205]
		H_2O/ 982 ← 981	$2.851 \cdot 10^{-24}$	1–2	na		
II	4	$H^{35}Cl$/ R3	$1.250 \cdot 10^{-20}$	2–5	$1.253 \cdot 10^{-20}$	0.8	TDLAS [106]
	5	CH_4/R(3)-triplet[28]	$3.170 \cdot 10^{-22}$	>20	$2.999 \cdot 10^{-21}$	< 1	TDLAS [208]
	6	C_2H_2/ P 23e	$2.214 \cdot 10^{-21}$	1–2	$2.220 \cdot 10^{-21}$	0.5	FTIR [209]
	7	OH/ PP 8.5ff	$2.183 \cdot 10^{-23}$	10–20	na		
	8	H_2O/ 202 ← 303	$1.500 \cdot 10^{-20}$	5–10	$1.470 \cdot 10^{-20}$	1.8	TDLAS [207]
		H_2O/ 660 ← 661[28]	$7.908 \cdot 10^{-22}$	10–20	$7.890 \cdot 10^{-22}$	1.7	TDLAS [207]
III	9	O_2/ R5Q6	$8.059 \cdot 10^{-24}$	<1			
		O_2/ R7R7	$7.380 \cdot 10^{-24}$	<1			

Temperature coefficient n

The temperature broadening coefficient by air n_{air} is known for the 202←303 H_2O and the R(19) CO transition as well as the R(3)-triplet of CH_4 [206,208,210–212]. The references also list the temperature self-broadening coefficient for the CO, CH_4 and the R(18)- CO_2 transitions. For CO_2, the effects of H_2O, N_2 and O_2 are known as well as those by H_2O, CO_2, N_2 and CO on CH_4. The temperature-broadening of the CO transition is listed for CO.

Pressure broadening γ and shift δ

The foreign-pressure broadening coefficients of the 202←303 water vapor line at 7181.16 cm^{-1} were measured for O_2, N_2 and air [206,213–215]. Experimental data on the pressure shift is only available for air [215]. For the 313←212 absorption line at 5393.65 cm^{-1}, the self-induced pressure broadening was experimentally measured using FTIR spectroscopy [216], but no experimental data is available on foreign pressure broadening and pressure shift coefficients. The self, O_2-, N_2- and air-induced pressure broadening coefficients of the R(18) line of CO_2 at 4991.26 cm^{-1} have been calculated, but not experimentally verified [48, 49]. Régalia-Jarlo et al. [45] measured the self-induced pressure broadening using a FTIR-spectrometer. Similar to the H_2O lines, no experimental data is available on the pressure shift.

The self- and N_2-broadening of the O_2 A-band was measured by Brown et al. [217]. The R(19)-transition of CO is comprehensively investigated, with available data on the self-, CO_2-, N_2-, O_2-, HCl- and air-collisional broadening as well as the self-, CO_2-, N_2-, and air-pressure shift [210,211,218–222]. As these include the major species of oxy-fuel combustion, there is no need to investigate this transition in further depth.

Self-, CO_2-, N_2-, air-, and O_2-induced pressure shifts in combination with the corresponding pressure broadening coefficients of the P(23) line of C_2H_2 were measured by Arteaga et al. [223] and Minutolo et al. [224]. However, the experiments were performed at relatively low pressures and therefore need

---

[26] cm^{-1}/(molec $\cdot$ cm^{-2})

[27] na: not accessible

[28] Since HITRAN2008, these absorption lines of CH_4 and H_2O are separated into three (respectively two, 660←661 and 661←660) independent transition at the same spectral position and almost identical line shape coefficients. These lines can not be separated at ambient pressure. Hence, combined intensity, broadening and shift coefficients will be given in the following, which is often done in literature [206–208].

validation at higher pressures. For the CH_4 R(3) line triplet, pressure broadening coefficients by CO, N_2, O_2, CO_2, air and CH_4 itself were measured, as well as the pressure shifts by air and N_2. Hence, these coefficients can be used for a comparison with the values determined in this work.

For the R(3) line of HCl, Orthwein et al. [106] and Li et al. [225,226] determined the He, N_2 and O_2 pressure broadening coefficients.

In summary, it can be stated that with the exception of the CO transition, pressure broadening and shift coefficients other than by the species itself or air are only sporadically known, leaving a large scientific gap. For the transitions listed in table 3.4, the collisional broadening and pressure shift coefficients by the listed species were measured and will be described in the next subsection.

Table 3.4: Collisional broadening and pressure shift coefficients by various combustion-relevant species to be measured.

System	#	target species/transition	collisional broadening γ and pressure shift δ gases
I	2	CO_2/ R39e	self, N_2, O_2, air, C_2H_2, CH_4
	3	H_2O/ 212 ← 313	CO_2, N_2, O_2, air, C_2H_2, CH_4
II	5	CH_4/R(3)-triplet	self, CO_2, N_2, C_2H_2
	6	C_2H_2/ P 23e	self, CO_2, N_2, CH_4
	8	H_2O/ 202 ← 303	CO_2, N_2, O_2, air

3.5.2 Experimental Setup

The self-induced collisional broadening and pressure shift can be obtained via a pressure variation of a target gas sample. Thereby, the foreign broadening and shift in eqn. 35 and eqn. 36 can be neglected and the line width and position change linearly with pressure. The foreign-induced collisional broadening and pressure shift are measured independently for each species via pressure variation of a mixture of mainly one foreign gas and a small amount of the absorbing target gas. Due to the small amount of target gas, the impact of self-broadening and shift are negligible and the line width and position change again linearly with pressure.

The experiments were conducted in two separate single-pass absorption cells. The first cell (SB) is specifically designed to measure the self-induced pressure broadening and shift (fig. 3.13). In pure target gas, the absorption is high, hence the cell is relatively short. Laser light is guided into the cell through a fiber, collimated inside and propagates 20 mm through the cell onto a photodiode.

The second cell is designed to measure the foreign-induced pressure broadening and shift (FB-cell). For each experiment, the target gas is mixed with a single foreign gas in predominant concentration (typically 98–99.5%$_{\text{Vol.}}$), leading to a negligible self-broadening compared to the foreign-broadening. Hence, the absorption of the target gas is very low. To account for this effect, the FB cell is significantly longer than the SB cell. The laser light is fiber-coupled into the cell and divergently exiting the angled fiber tip. An adjustable off-axis parabolic mirror guides the light onto a detector, yielding an effective absorption length of 223 mm. For the fibers and detector wires, sealed feed-throughs were installed into the cell walls.

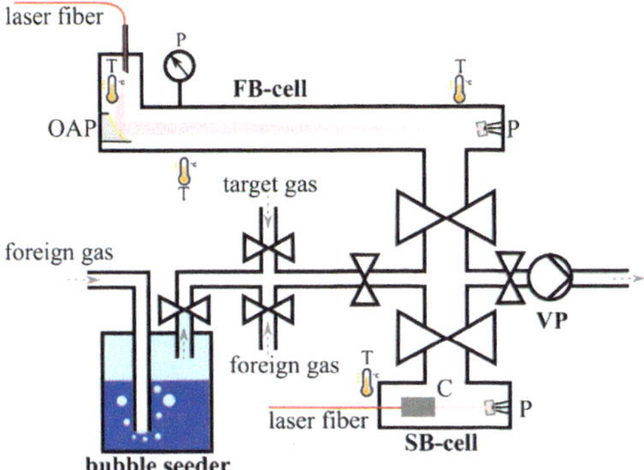

Fig. 3.13: Measurement cells and environment. FB-cell: Foreign broadening/shift cell; SB-cell: Self broadening/shift cell; VP: Vacuum pumps; T: location of thermocouples; P: location of pressure gauge; P: Photodiode; C: Collimator; OAP: Off-axis parabolic mirror.

Prior to each measurement, the cells were evacuated to <1 Pa using a HiPace turbomolecular pump in combination with a MVP-006-4 diaphragm pump (Pfeiffer Vaccum). The pressure was measured using the pressure transducers described in section 3.4. The coefficients of all lines were determined at reference temperature (296 K) by stabilizing the cell temperature through air-conditioning. Multiple calibrated thermocouples (type K, 0.2 K measurement uncertainty) along the cells were used to measure the gas temperature and temperature homogeneity. The cells were filled with gas through a set of chambers and valves. Target and foreign gas were mixed within the cell and given a settling time to achieve a mixing homogenization and thermal equilibrium with the walls. The target-to-foreign gas ratio was adjusted by injecting the target gas up to a desired partial pressure and subsequently filling with foreign gas to the desired total pressure. The turbulence inside the cell introduced by this procedure ensures that the gases were well mixed. The target concentrations were double-checked using the TDLAS measurements. For water vapor as target gas, the foreign gas was moisturized using a bubble-seeder. The saturation pressure of water at ambient temperature (~3.5 kPa) impedes to use water vapor as foreign gas. For the experiments, the 2004 nm CO_2, the 1854 and 1392 nm H_2O, the 1654 nm CH_4 and the 1539 nm C_2H_2 lasers listed in table 3.1 were used. In the measurements in this chapter, the modulation rate of all lasers was set to 39 Hz to achieve a high spectral resolution in the range of 10^{-3} cm^{-1}, allowing to sufficiently resolve the absorption lines in both the low (5 kPa) and high pressure range (180 kPa).

3.5.3 Evaluation and Results

Integration time and absorption line profiles

To determine broadening and shift coefficients, 1000 absorption spectra each were recorded at various pressures between 5 and 180 kPa. In the fitted model, the absorption line strength, the line center position and Lorentz-width of the absorption line were independent variation parameters. Due to measurement noise mainly from electrical white noise, each measured spectrum yields a slightly different position and width. Hence, averaging of the data is necessary.

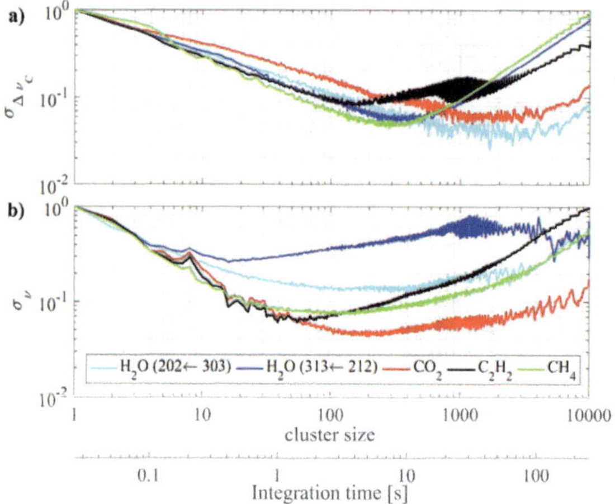

Fig. 3.14: Allan-Werle variances of the absorption lines influenced by N$_2$ foreign gas. a) pressure broadening b) line position

However, the averaging time should not exceed a certain number of spectra, as otherwise systematic errors are included. Here, the Allan-Werle variance was used to determine the optimal cluster size or integration time for averaging. Fig. 3.14 shows the measured Allan-Werle variances of the five absorption lines for 50000 consecutive spectra. During the measurements, the lines were broadened by N$_2$ as foreign gas. The total pressure of the measurements was ~80 kPa and the concentration of the target species $< 2\%$. Fig. 3.14a shows the variance of collisional broadening, which is initially decreasing due to the elimination of white noise. For cluster sizes >300 (integration time >8 s) the variance is minimal. For higher cluster sizes, it is increasing. Here, minor drifts come into play, probably a temperature or pressure drift within the cell, as the drift is roughly independent of the spectrometer used. Therefore, for the determination of broadening coefficients, 300 absorption profiles were averaged. The Allan-Werle variance of the line position, however, seems to be spectrometer-dependent. Fig. 3.14b shows a significantly earlier drift for measurements at the 313←212 H$_2$O line compared to CO$_2$. As a drift in the measured line position is affected by drifts in the diode laser temperature, they probably stem from fluctuations in the laser cooling. In fact, the 1854 nm DM laser in the 313←212 H$_2$O spectrometer is the only one exhibiting an early drift, whereas the other spectrometers, which rely on DFB and VCSEL laser technology, drift at lower frequencies. For the DM laser, the variance is minimal for a cluster size of 20 (integration time 0.5 s), whereas it is minimal for a cluster size of 100 for the other spectrometers. Hence, to determine the shift coefficients, the integration times were chosen accordingly.

Fig. 3.15 shows measured absorption line profiles for various pressures. In fig. 3.15a, the foreign broadening of 313←212 H$_2$O and adjacent absorption lines by N$_2$ as foreign gas are shown for pressure levels between 5 kPa and 180 kPa. With increasing pressure, the lines are spectrally broadened and begin to overlap. In order to separate the influence of the adjacent lines, the whole spectrum needs to be fitted. The residual shows the difference between the measured spectrum at 180 kPa and the model. Due to the comparatively low variance of $4.2 \cdot 10^{-4}$ and the high peak line strength of 0.23, the signal to noise ratio (SNR= $\alpha_{Peak}/\sigma_{Residual}$) lies in the order of 570, which is typical for this spectrum and results in a low fit uncertainty. Here, the residual is dominated by line profile deviations due to the complex structure of the spectrum, where some line positions and broadenings of the background are not known accurately.

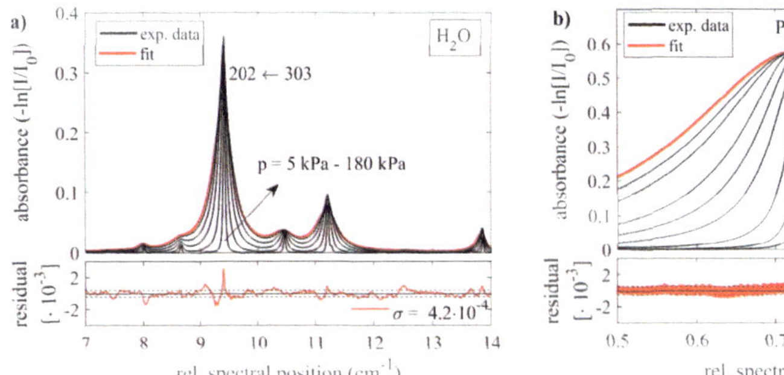
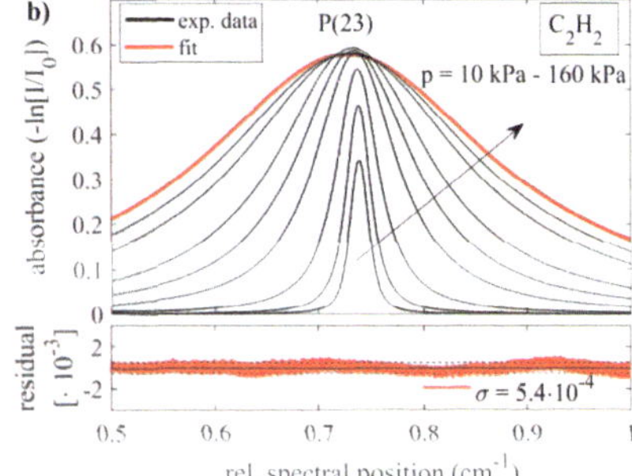

Fig. 3.15: Exemplary absorption line profiles for various pressures. The residual is the difference between the measured data and the modeled fit for the highest pressure level. a) Measured line profiles of H_2O mixed with N_2 as foreign species. b) Measured line profiles of C_2H_2 under self-broadening.

Fig. 3.15b shows the absorption profile of the C_2H_2 P(23) line under self-broadening with pressure levels between 10 kPa and 160 kPa. Here, the spectrum is dominated by one isolated absorption line and is therefore significantly less complex. Besides the broadening of the line, a small shift of the line peak towards lower wavelengths with increasing pressure is evident, which is the pressure shift. Due to the simple profile, the model and the experimental data agree better and the SNR exceeds 10^4. Only minor deviations are evident, the residual is dominated by electrical noise, probably from the detector and amplifier. Due to the high SNR in both examples it can be concluded that a very good agreement between model and experimental data is achieved and the uncertainties corresponding to errors in the profile are low.

Error Analysis

Multiple sources of error affect the uncertainty of the broadening and shift coefficients, such as errors in pressure and temperature, purity of the gases, quality of the absorption profile fit, uncertainty of the laser tuning and the linear regression. The uncertainty in absorption path length, which is an important factor in TDLAS concentration measurements, does not influence the uncertainty of the broadening and shift coefficients directly, as these are independent of the path length. Only the measurement of the target gas partial pressure using TDLAS is influenced by this parameter. However, the indirect contribution to the broadening and shift parameters can be neglected as it is small compared to other uncertainties.

The laser tuning coefficient $\partial v / \partial t$ is responsible for the accuracy of the spectral position and thus influences the broadening and shift coefficients. The uncertainty of this factor is dominated by the uncertainty of the optical path length in the etalon. Remaining impurities of the target and foreign gases affect both the broadening and shift coefficients. The purity of the gases ranges from 99.5% for CH_4, O_2, air and water vapor to 99.9999% for N_2. While the uncertainty of the spectral fit through the SNR is described above, the error in the regression of the measured coefficients with pressure (Δm) will be discussed later on in this subsection.

Fig. 3.16 visualizes the relative contributions of the different sources of error to the uncertainty of all broadening and shift coefficients. Each of the uncertainty sources is relevant during at least a few of the measurements. Hence, the error cannot be significantly reduced by improving one factor only, which makes a further reduction of the uncertainty complex. A more detailed analysis of each contribution can be found in Bürkle et al. [E2].

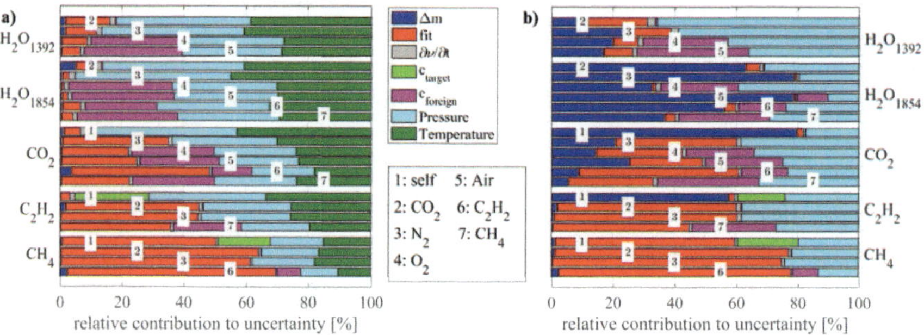

Fig. 3.16: Relative influences of the uncertainty sources on the overall error a) self- and foreign broadening b) self- and foreign shift

It was not possible to determine the C_2H_2 and CH_4 broadening coefficients of the 202←303 water vapor absorption line, as minor absorption bands of the foreign species overlap with this absorption line. Here, the different spectra could not be isolated sufficiently due to a low quality of the available spectral data for the foreign species. The foreign broadening and shift coefficients of C_2H_2 and CH_4 in O_2 and air were not measured due to an explosion hazard. The foreign broadening by water vapor was not measured due to the low saturation pressure at ambient temperature. All uncertainties account for total errors in the range of 0.7–1.5%$_{rel.}$ for the pressure broadening coefficients and 0.6–1.6%$_{rel.}$ for the drift coefficients (table 3.5 and table 3.6).

Self- and foreign pressure broadening

Fig. 3.17 exemplarily shows the Lorentz half-width of the R(18) CO_2 absorption line for various pressures and collisional partners. During measurements of the foreign broadening, the effects of self-broadening are subtracted to reduce systematic errors, even though the influence was comparatively small. For these calculations, the target gas concentration measured by TDLAS was used. The linear fit in fig. 3.17 was independent of the zero pressure/zero broadening point to check for offsets in the pressure sensors, which were below 0.8 kPa.

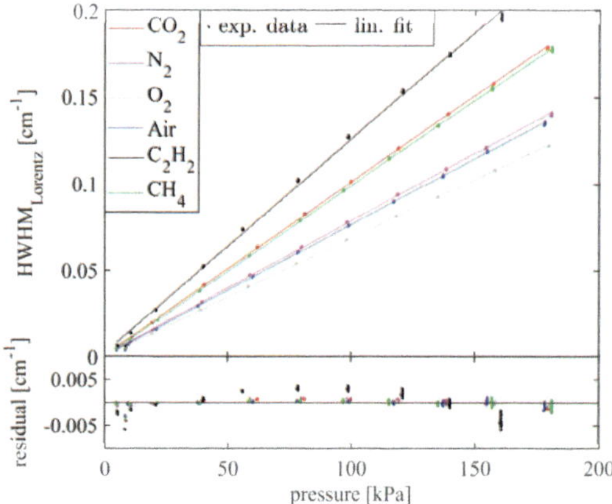

Fig. 3.17: Self- and foreign pressure-broadening of CO_2 and comparison with linear fit

The slope of the R(18) CO_2 absorption line is linear but species dependent. C_2H_2 is broadening the absorption line twice as strong as O_2. For a better comparison with the coefficient provided in HITRAN, the air broadening coefficient is measured directly and not only calculated from the N_2 and O_2 coefficients. The linearity of the data is very good with R^2 values >0.9999. Hence, the absolute uncertainty of the slope Δm, estimated by the standard deviation of the slope [227], is low.

The self-broadening coefficient of the R(18) absorption line was determined to be $\gamma_{self} = (0.1011 \pm 0.00073)$ cm^{-1}/atm, which greatly matches the HITRAN2016 value ($\gamma_{self} = (0.1 \pm 0.02)$ cm^{-1}/atm) [88]. Here, a significant improvement of the uncertainty was achieved compared to the literature. The experimentally measured coefficient by Régalia-Jarlo et al. [228] undercuts the value measured in this work by -5%, whereas the calculated coefficient by Rosenmann et al. [212] exceeds it by +5%.

The measured air-broadening coefficient is consistent with the measured O_2 and N_2 broadenings, when taking the relative amounts of the gases in air into account. However, it slightly exceeds the uncertainty of the coefficient provided in HITRAN2016 ($\gamma_{Air} = (0.0735 \pm 0.0015)$ cm^{-1}/atm) and Toth et al. [229]. The N_2-broadening coefficient slightly undercuts the error margins of the value provided by Rosenmann et al. [212], whereas the O_2-coefficient is in perfect agreement with the value provided by the same publication.

Table 3.5.: Self- and foreign pressure-broadening coefficients, uncertainties, comparison to literature and uncertainty of literature value.

Line pos.		γ [cm⁻¹/atm]	$\Delta\gamma_{abs.}$ [cm⁻¹/atm]	$\Delta\gamma_{rel.}$ [%]	$1-R^2$	HITRAN 2016 [cm⁻¹/atm]	$\Delta\gamma_{rel.}$ HITRAN [%]	Other sources [cm⁻¹/atm] [source]	Err. [%]
H_2O 1392 (7181.16 cm⁻¹)	γ_{CO2}	**0.1720**	±0.00131	0.76	$7.7 \cdot 10^{-5}$				
	γ_{N2}	**0.1082**	±0.00080	0.74	$2.5 \cdot 10^{-5}$			0.12 [214] 0.1138 [213]	
	γ_{O2}	**0.0625**	±0.00056	0.90	$4.4 \cdot 10^{-5}$			0.07 [214] 0.0644 [213]	
	γ_{Air}	**0.0978**	±0.00086	0.89	$2.9 \cdot 10^{-5}$	0.1008	10	0.1034 [213] 0.0969 [230] 0.09693 [231] 0.0995 [215]	2 2 1
H_2O 1854 (5393.65 cm⁻¹)	γ_{CO2}	**0.1323**	±0.00098	0.74	$1.9 \cdot 10^{-5}$				
	γ_{N2}	**0.1034**	±0.00073	0.71	$1.1 \cdot 10^{-5}$				
	γ_{O2}	**0.0567**	±0.00049	0.86	$1.4 \cdot 10^{-5}$				
	γ_{Air}	**0.0944**	±0.00082	0.87	$2.7 \cdot 10^{-5}$	0.095	1		
	γ_{C2H2}	**0.2007**	±0.00167	0.83	$2.0 \cdot 10^{-5}$				
	γ_{CH4}	**0.0879**	±0.00077	0.87	$3.0 \cdot 10^{-5}$				
CO_2 (4991.26 cm⁻¹)	γ_{self}	**0.1011**	±0.00073	0.72	$1.6 \cdot 10^{-5}$	0.1	2	0.0964 [228] 0.106 [212]	
	γ_{N2}	**0.0782**	±0.00067	0.86	$4.0 \cdot 10^{-5}$			0.0796 [212]	
	γ_{O2}	**0.0683**	±0.00066	0.97	$3.1 \cdot 10^{-5}$			0.0683 [212]	
	γ_{Air}	**0.0765**	±0.00076	0.98	$5.6 \cdot 10^{-5}$	0.0735	2	0.0738 [229]	1
	γ_{C2H2}	**0.1244**	±0.00139	1.11	$9.2 \cdot 10^{-5}$				
	γ_{CH4}	**0.0990**	±0.00096	0.97	$4.8 \cdot 10^{-5}$				
C_2H_2 (6495.91 cm⁻¹)	γ_{self}	**0.1109**	±0.00090	0.82	$5.6 \cdot 10^{-6}$	0.116	5	0.1 [224]	1
	γ_{CO2}	**0.0728**	±0.00068	0.94	$9.7 \cdot 10^{-5}$			0.0702 [224]	1
	γ_{N2}	**0.0679**	±0.00063	0.93	$2.7 \cdot 10^{-5}$			0.0672 [223] 0.0697 [224]	2 2
	γ_{CH4}	**0.0837**	±0.00090	1.07	$1.8 \cdot 10^{-5}$				
CH_4 (6046.95 cm⁻¹)	γ_{self}	**0.0984**	±0.00120	1.21	$5.2 \cdot 10^{-5}$	0.079	>20	0.0766 [208] 0.085 [232]	2 5
	γ_{CO2}	**0.0805**	±0.00093	1.16	$2.1 \cdot 10^{-5}$			0.0746 [208]	2
	γ_{N2}	**0.0620**	±0.00096	1.12	$4.1 \cdot 10^{-5}$			0.056 [208] 0.068 [232]	1 6
	γ_{C2H2}	**0.0981**	±0.00143	1.45	$1.1 \cdot 10^{-3}$				

Table 3.5 summarizes the pressure broadening coefficients for the five investigated absorption lines and provides the absolute and relative errors of the measurements and the R^2 values of the linear fit. In addition, the available HITRAN2016 and most important literature values in combination with available uncertainties are given. As a general trend, the C_2H_2 broadening is always the strongest and the O_2 broadening the weakest. In between, the order of the CO_2, N_2 and CH_4 broadening coefficients depends on the target species. For the $202\leftarrow303$ absorption line of H_2O, low uncertainties of less than 1% are achieved. The influence of different foreign gases is similar to CO_2 as target gas, with CO_2 showing a strong broadening, whereas N_2 and O_2 are weaker broadening partners. The air-broadening coefficient is again confirmed by the O_2 and N_2 coefficients and greatly matches the HITRAN2016 value and the ones given in [213,215,231]. However, the uncertainty is significantly reduced compared to [213,231] and is even lower than that of [215]. The N_2 and O_2 coefficients agree well with literature values, where no uncertainties are provided. Similar to this H_2O absorption line, measurements at the $313\leftarrow212$ line led to very low uncertainties. Here, only the HITRAN2016 value for the air-broadening was found ($\gamma_{Air} =$ (0.095± 0.005) cm^{-1}/atm). The measured value agrees very well with this coefficient. For the other broadening coefficients of the $313\leftarrow212$ line, no literature values were found. The measured self-broadening of C_2H_2 agrees well with HITRAN2016 but has significantly lower uncertainties. In addition, literature values for the CO_2 and N_2 broadening were reproduced.

The measured self-broadening of the CH_4 R(3) line-triplet is highly different from the HITRAN2016 coefficient. However, the measured value is still within the uncertainty of the latter, as HITRAN2016 has large uncertainties for this quantity. Compared to other literature values, the measured foreign-broadening is sporadically outside the range of uncertainty, but the literature values disagree among each other as well. For example, the measured N_2-broadening coefficient exceeds the value measured by Gharavi and Buckley [208] by 10%, but undercuts the value by Varansi et al. [232] by the same amount.

In general, the measured coefficients agree very well with literature values where available, but the uncertainties were often significantly reduced. In approximately half of the cases, the coefficients were not known at all prior to the measurements of this work.

Self- and foreign pressure shift

Fig. 3.18 exemplarily shows the relative line position (normalized with the origin of the linear fit) of the R(18) CO_2 absorption line for various pressures and collisional partners. The absorption line linearly moves towards smaller wavenumbers with increasing pressure. However, the slope is dependent on the foreign species, as the absorption line is shifted nearly twice as far from collision with CH_4 as with N_2. The linearity of the data is not as good as for the pressure broadening with R^2 values ranging from >0.995 to >0.99999 (table 3.6) which, however, is still very high. Fig. 3.18 also shows that the residual is originating from two factors: scattering of the values around the mean and a systematic deviation from linearity. The scattering probably stems from the fitting of the absorption line to the spectral data. As the absorption lines, in particular for high pressures, are broad, the line center cannot be identified perfectly by the algorithm, as electrical noise on the detector signal leads to a reduced precision in line position. For the deviation from linearity, two reasons are possible:

1. *Temperature fluctuations of the laser:* If the laser temperature is not controlled perfectly, the laser central wavelength is slowly drifting, and thus the absorption line seems to virtually move. As the temperature control is rather slow (~min), this effect could lead to systematic deviation from linearity in consecutive measurement points. One option to eliminate effects of temperature fluctuations is to perform a line-lock by splitting the laser into two beams, with the first one measuring within the cell under varying pressure and the second one measuring within a static absorption cell. This procedure was not done during the experiments and thus contributes to the overall uncertainty of <2% by a fraction of 0.1–0.8%.

2. *Errors from the fitting algorithm*: At high pressures, the laser baseline cannot be resolved perfectly due to the significantly broadened absorption lines compared to the laser tuning range. Then, an interaction between the fitting of the baseline and the absorption line can lead to a falsely fitted line position.

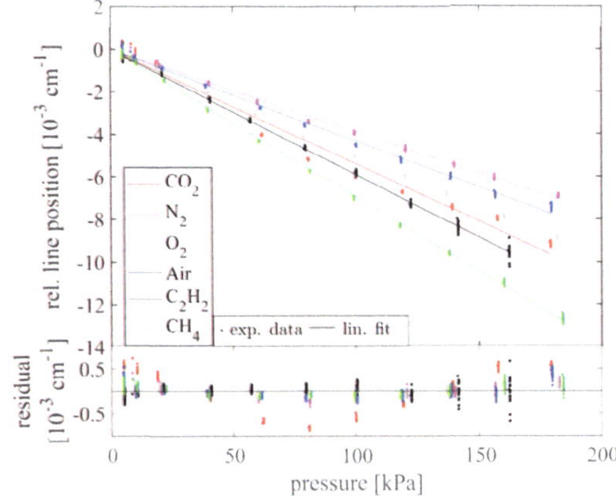

Fig. 3.18: Self- and foreign pressure-shift of CO_2 and comparison with linear fit

Literature values on the pressure shift are sparse, as the shift is usually in the same magnitude as the spectral resolution of the commonly used FTIR spectrometers. HITRAN2016 provides the air-induced pressure shift of the R(18) CO_2 line (δ_{Air} = -5.39) with a high uncertainty of 20%. Toth et al. [229] determined the value with a lower error of only 10%. The shift coefficient of δ_{Air}=-4.38·10^{-3} cm^{-1}/atm measured in this work is within the error margins of both values. It exceeds the coefficient by Toth et al., undercuts the HITRAN value and has a significantly lower uncertainty of only 1%. For the other pressure shift coefficients of the CO_2 R(18) absorption line, no literature values were found.

Table 3.6 lists the measured self- and foreign-induced pressure shifts of the five absorption lines and provides the absolute and relative errors of the measurements in combination with the R^2 values of the linear fit. In addition, the available HITRAN2016 and most important literature values are given. The HITRAN2016 air-shift coefficient of the 202←303 absorption line of water vapor is relatively close to the measured value of δ_{Air} = -13.43·10^{-3} cm^{-1}/atm, considering the uncertainty of the HITRAN value (labeled: default). The measured value is also very close to the one provided by Hunsman et al. [215] and has a significantly lower uncertainty of less than 1%. Similar to the 202←303 absorption line of water vapor, the air-shift of the 313←212 absorption line provided in HITRAN2016 with unknown uncertainty, δ_{Air} = -2.99·10^{-3} cm^{-1}/atm, is rather close to the measured value of δ_{Air} = -3.41·10^{-3} cm^{-1}/atm. The foreign-shift of the 202←303 absorption line is positive in some cases, which means the absorption line is moving towards larger wavenumbers with increasing pressure. The measured position of the 313←212 absorption line shows increased deviations from linearity, probably due to an increased temperature fluctuation of the laser's central wavelength, as described above.

While the measured C_2H_2 self- and N_2-shifts are within the uncertainty of the values provided by Minutolo et al. and Arteaga et al. [223,224], respectively, the N_2-shift is slightly outside the uncertainty range of the coefficient determined by Minutolo et al. The CO_2-shift measured by the same group is also far off the value measured in this work. For the CH_4 R(3)-triplet, one combined pressure shift is determined. The N_2-shift of this line in good agreement with the value provided by Dufour et al. [233].

Table 3.6: Self- and foreign pressure-shift coefficients, uncertainties and comparison to literature.

Line pos.		δ [$\cdot 10^{-3}$ cm^{-1}/atm]	$\Delta\delta_{abs.}$ [$\cdot 10^{-3}$ cm^{-1}/atm]	$\Delta\delta_{rel.}$ [%]	$1-R^2$	HITRAN 2016 [cm^{-1}/atm]	$\Delta\delta_{rel.}$ HITRAN [%]	Other sources [$\cdot 10^{-3}$ cm^{-1}/atm] [source]	Err. [%]
H_2O 1392, 7181.16 cm^{-1}	δ_{CO2}	-22.21	±0.137	0.62	$7.2 \cdot 10^{-4}$				
	δ_{N2}	-12.82	±0.084	0.65	$2.2 \cdot 10^{-3}$				
	δ_{O2}	-18.44	±0.141	0.76	$1.7 \cdot 10^{-3}$				
	δ_{Air}	-13.43	±0.112	0.83	$2.5 \cdot 10^{-3}$	-12.509	default	-13 [215]	7.7
H_2O 1854, 5393.65 cm^{-1}	δ_{CO2}	4.67	±0.042	0.90	$1.0 \cdot 10^{-2}$				
	δ_{N2}	-2.24	±0.025	1.13	$2.3 \cdot 10^{-2}$				
	δ_{O2}	-7.50	±0.060	0.80	$4.0 \cdot 10^{-3}$				
	δ_{Air}	-3.41	±0.053	1.56	$3.8 \cdot 10^{-2}$	-2.999	unavailable		
	δ_{C2H2}	4.05	±0.041	1.02	$1.1 \cdot 10^{-2}$				
	δ_{CH4}	-6.19	±0.057	0.92	$5.4 \cdot 10^{-3}$				
CO_2 4991.26 cm^{-1}	δ_{self}	-5.45	±0.065	1.19	$3.8 \cdot 10^{-3}$				
	δ_{N2}	-3.92	±0.032	0.81	$1.1 \cdot 10^{-3}$				
	δ_{O2}	-5.00	±0.043	0.85	$7.7 \cdot 10^{-4}$				
	δ_{Air}	-4.38	±0.044	1.00	$1.8 \cdot 10^{-3}$	-5.393	20	-4.8 [229]	10
	δ_{C2H2}	-5.97	±0.062	1.03	$2.0 \cdot 10^{-3}$				
	δ_{CH4}	-7.03	±0.061	0.87	$3.7 \cdot 10^{-4}$				
C_2H_2 6495.91 cm^{-1}	δ_{self}	-6.81	±0.069	1.02	$5.6 \cdot 10^{-3}$			-6.1 [224]	20
	δ_{CO2}	-11.52	±0.093	0.81	$1.3 \cdot 10^{-4}$			-9.4 [224]	3
	δ_{N2}	-8.67	±0.070	0.81	$8.0 \cdot 10^{-4}$			-8.6 [223]	20
								-9.0 [224]	2
	δ_{CH4}	-14.39	±0.138	0.96	$8.0 \cdot 10^{-5}$				
CH_4 6046.95 cm^{-1}	δ_{self}	-15.65	±0.176	1.12	$8.8 \cdot 10^{-5}$				
	δ_{CO2}	-11.77	±0.125	1.07	$8.8 \cdot 10^{-6}$				
	δ_{N2}	-9.69	±0.098	1.01	$1.6 \cdot 10^{-5}$			-9.96 [233]	0.1
	δ_{C2H2}	-13.43	±0.184	1.37	$5.3 \cdot 10^{-4}$				

In summary, the measured coefficients usually lie within the uncertainty of the few available literature values, as uncertainties of the latter are usually very high. For these parameters, the uncertainty was significantly reduced. The majority of the shift coefficients was reported for the first time.

3.6 Conclusions

In this chapter, an optical measurement system for measuring the thermochemical state of oxy-fuel combustion systems was developed. The system consists of three subsystems based on absorption spectroscopy with a total of 9 lasers, which are used to scan 13 absorption lines of 8 combustion-relevant species. To the best of the author's knowledge, this is the largest in-situ TDLAS system with respect to the number of investigated species and used lasers in literature. Basic concepts for the optical setup have been developed which allow a versatile adaption to the different systems under investigation. The data acquisition setup and the principle uncertainty analysis have been described. The relevant absorption lines have been carefully selected and their absorption coefficients, which are necessary for a model description of the absorption line, were inferred from literature. In many cases, experimental measurements for an improvement of the coefficients were imperative and were performed in two specially designed measurement cells. These experiments led to a unique dataset of high accuracy, which is considered to be included into the next HITRAN edition by the HITRAN editors. [29]

In the next chapters, this measurement system will be verified on a well-investigated flame at first. Afterwards, it is used to investigate several oxy-fuel combustion systems.

---

[29] Personal communication with Iouli Gordon, Harvard-Smithsonian Center for Astrophysics, Cambridge

4 Laminar Oxy-Gas Combustion: The Wolfhard-Parker Burner

For potential oxy-fuel combustion processes, such as gas turbines or industrial solid fuel furnaces, the combustion of gaseous components is essential. Both practical applications partially employ non-premixed combustion. Gas turbine injectors introduce the fuel into a stream of oxidizer without prior mixing. In solid fuel combustion furnaces, the devolatilization of solid fuel particles create a fuel-rich region around the particles, which itself is surrounded by the oxidizer-rich combustion atmosphere. While both applications operate in the turbulent flow regime, turbulent combustion is complex to investigate and the interpretation of the results difficult, due to the small time- and length-scales involved as well as the complexity of the flow. Therefore, fundamental investigations of the processes involved in oxy-gas combustion and their differences to air-blown combustion are often conducted in generic laminar flow burners, mainly of counter-flow type [13,234–240], due to their simplicity, stability and absence of walls. However, the effects can also be studied in co-flow burners such as the *Wolfhard-Parker* burner (WHP), whose dimensionality can be reduced to a 2D-Cartesian description of one burner half-plane. This characteristic allows a reduced complexity and effort in numerical simulations as well as the application of path-integrating measurement techniques. Additionally, the air-blown operation of this burner has been subject to various investigations of the flow-field and the thermochemistry, which allows to use this operating condition for validation and error analysis of the developed measurement system.

For the oxy-gas combustion, a new operating condition of this burner with Oxy30 combustion atmosphere (30 %$_{Vol.}$ O$_2$/70 %$_{Vol.}$ CO$_2$) has been defined in addition to the air-blown flame. To simplify typical gas turbine fuels and the volatiles of solid fuel, methane (CH$_4$) has been used as fuel in both operating conditions. Profiles of the thermochemical state of both operating conditions are measured for different axial distances to the burner.

The aims of the investigations presented in this chapter are:

- to validate the measurement system described in the previous chapter using a well-known burner and corresponding reference data,

- to compare profiles of the thermochemical state (gas temperature, mole fractions of H$_2$O, CH$_4$, CO and OH) and the flame position of a laminar, non-premixed oxy-gas flame with an air-blown methane flame of the same type,

- to generate experimental validation data for numerical models, in particular for the significant radiative heat loss in oxy-gas environments.

The features and design of the WHP burner and the operating conditions are described in detail first, followed by an outline of the optical setup and the spectrometers used. The uncertainties of the measurements are discussed hereinafter, followed by a presentation of the measurements and discussion of the results. Parts of this chapter have already been published by the author in [E5].

4.1 Experimental Setup

4.1.1 Wolfhard-Parker burner

The Wolfhard-Parker co-flow burner was originally described by Wolfhard et al. [241] for ethylene-oxygen flames. In this work, it is employed in the improved design for methane-air flames by Smyth et.

al. [242,243] and Wagner et al. [97], who stabilized the flame by introducing a second oxidizer slot. As shown in fig. 4.1, the burner consists of three parallel, rectangular slots of 40 mm length, with the central slot ejecting methane and the outer ones providing an oxidizing combustion atmosphere (either air or Oxy30). The flow is homogenized inside the burner using a settling chamber filled with silica beads and a copper wire mesh on top of it. The central fuel slot has a width of 8 mm; the exterior oxidizer slots span 16 mm each. Copper walls of 1 mm thickness separate the slots. The non-premixed gas flows atop these walls stabilize two symmetrical flame-sheets, creating a homogeneous, 41 mm long absorption path along the slot axis (y-direction) for the laser. At the end of the fuel slot, the burner is equipped with two additional slots (design by Wagner et al. [97]) of 7.5 mm x 16 mm for purging with N_2 during air operation and CO_2 during combustion within an oxy-fuel atmosphere. These prevent edge flames at the slot tips as well as a stretch of the flow along the y-axis due to a shear between the flow and the ambient. The flame sheets are stabilized using a wire screen chimney above the burner as well as screens in the far-field around the burner to prevent disturbances by air drafts. The burner is automatically traversable in x-z direction using translation stages. Measured profiles are taken along the lateral (x-) axis at different heights (z-axis).

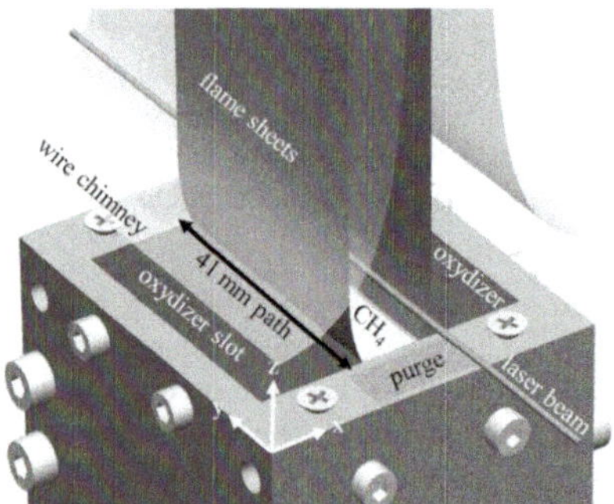

Fig. 4.1: Setup of the modified Wolfhard-Parker burner (left); photography the burner top surface, the wire chimney and the flame branches under air operation (right).

The operating condition of the burner under air atmosphere is identical to previous measurements [97,98,242–245] and summarized in table 4.1. To stabilize a flame under oxy-fuel conditions, the exit velocities of the oxidizer and purge flows have to be decreased by 50 % compared to the air-blown operation. This is mainly due to the reduced laminar flame speed. Other reasons are a lower oxygen diffusivity as well as an increased heat capacity. To decrease the strain rate, the exit velocity of the methane slot is reduced to a lesser degree (40 %) than that the of the oxidizer and purge slots. Generally, oxy-fuel combustion causes a twofold reduction of the stable operation range. In terms of flow velocity, the blow-off of the flame due to the reduced laminar flame speed needs to be avoided. For this reason, the exit flow velocity is reduced. This possibility is limited at lower values by the increased radiative heat transfer of the oxy-fuel flame to the copper flame holder, which needs to be kept low as it is quite complex to model. A stable operation was achieved within an Oxy30 combustion atmosphere, but not for lower oxygen contents. Table 4.1 summarizes the operating conditions. The Reynolds numbers demonstrate that the flows from all slots are in the laminar regime.

Table 4.1: Slot exit velocities for the two different operation cases

Case \ Slot	Central (CH₄)	Oxidizer	Purge (N₂ or CO₂)	Re-Number Central slot	Re-Number Oxidizer slot
Air	11 cm/s	22 cm/s	22 cm/s	53	230
OXY30	6.5 cm/s	11 cm/s	11 cm/s	31	51

For the air operation, the National Institute of Standards and Technology (NIST) [241,242] and Wagner et al. [97] published reference data of temperature and species concentration profiles for measurement heights of 7 mm, 9 mm and 11 mm above the burner. In the present work, the thermochemical state is measured at the same measurement heights and with the same lateral resolution. The lateral temperature profiles of the reference were obtained using radiation corrected thermocouples, concentration profiles of H_2O, CO, CH_4 and additional species using mass spectrometers (Smyth et al. [242]) and OH profiles using quantitative laser-induced fluorescence [243]. Additionally, the reference temperatures and CO-concentrations under conventional air-combustion were confirmed using 4.7 µm mid-infrared TDLAS by Miller et al. [246].

4.1.2 Optical Setup

To investigate the features of the laminar, non-premixed WHP flames, a high spatial resolution and precise alignment of the lasers are necessary. From the reference data ([243]), a flame thickness of ≈ 1 mm FWHM can be inferred. To sufficiently resolve this flame, a spatial resolution in the order of 0.2 mm is desired. The necessary displacement of the laser beam by the retroreflector impedes the usage of a single-sided setup. Hence, a single-pass setup was chosen at the cost of a reduced absorption length.

Fig. 4.2 shows the optical setup of the laser system. A fiber guides the laser light into a Grintec graded index lens, which collimates the beam. The lens is mounted into the central axis of an N_2-purged pipe, which is located above the purging slot of the WHP to prevent undesired absorption by ambient water vapor outside the flame. The exit velocity of the purging flow in the pipe is controlled in order to prevent a visible influence of the flow on the flame. The collimated laser beam propagates through the flame slots in parallel to the burner y-axis and is coupled into a 10 cm long 1000 µm multimode fiber (MM), which guides the light onto a photodiode (PD) with a 1 mm diameter of the active area. The multimode fiber isolates the laser beam from ambient water vapor absorption and simultaneously reduces thermal radiation from the flame onto the photodiode.

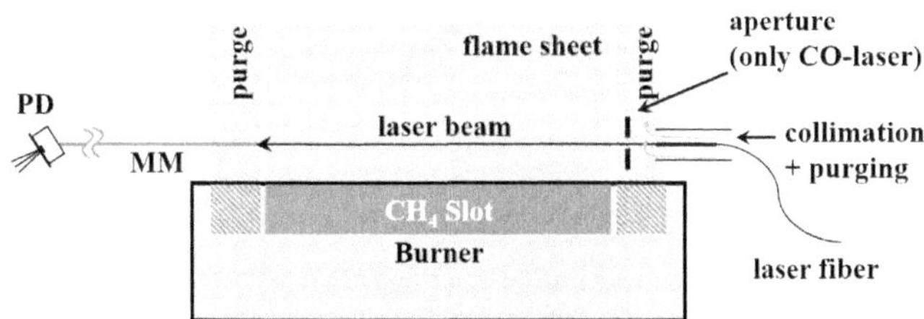

Fig. 4.2: Side view of the experimental setup with the WHP burner and applied laser spectrometers.

While the graded index lens is sufficient for collimating the lasers of System II, the attenuation by the lens and fiber are too high for the CO laser. Hence, for these lasers, a broadband lens collimator (Thorlabs) with the drawback of a wider beam waist (≈ 500 µm) had to be used. This beam diameter was reduced by a 200 µm aperture to achieve the desired spatial resolution.

From the thermochemical state reference data, estimations of the signal-to-noise ratios were deduced. From this analysis, four TDLAS-spectrometers were selected to measure the thermochemical state of the flame, as they promised sufficiently high signal-to-noise ratios. The spectrometers listed in table 4.2 were employed to measure the gas temperature and species concentrations of H_2O, CH_4, OH and CO. No use was made of the remaining lasers, as their respective lower detection limit exceeded the expected concentrations. While the H_2O, CH_4 and OH spectrometers were multiplexed, the steady nature of the flame allowed to operate the CO spectrometer independently and to thus simplify the handling. The laser-tuning scheme is described in section 3.2.

Table 4.2 Spectrometer and measured quantities at the WHP burner.

System	#	laser (central wavelength)/Target species	measured quantities
I	1	2314 nm CO	χ_{CO}
II	5	1653 nm CH$_4$	χ_{CH_4}
	7	1527 nm OH	χ_{OH}
	8	1392 nm H$_2$O	T_{H_2O}
			χ_{H_2O}

As the reaction zone thickness is very thin, a precise alignment of the laser with respect to the flame sheets and the burner surface, as well as knowledge of the laser-beam diameter, are crucial. To align the lasers and measure the laser profile in two directions (x- and z-axis), the *knife-edging* technique was used. In this technique, a blade is mounted either horizontally or vertically onto the front side of the burner. By moving the blade through the laser beam, the beam is partially blocked and the laser intensity on the photodiode reduced. Taking the spatial derivative of the detector signal leads to the laser intensity profile. Through a Gaussian profile fit, the beam half-width and central position are obtained. This process is repeated at the burner rear side both horizontally and vertically with the blade fixed at the same x- or z- positions to measure and iteratively correct the tilt of the laser with respect to the burner axis.

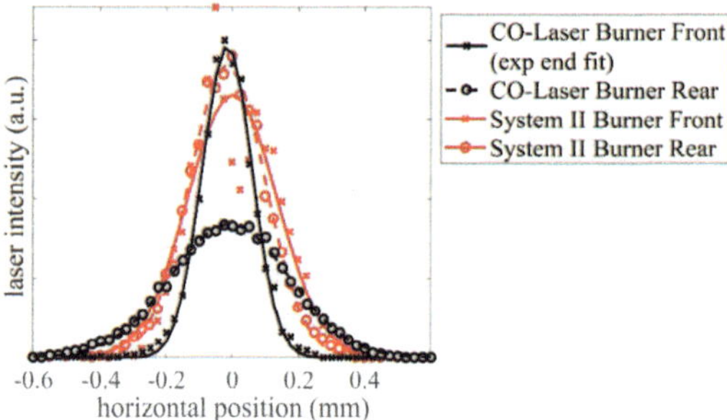

Fig. 4.3: Laser profile of System II and the CO laser in horizontal direction.

Fig. 4.3 exemplarily shows the horizontal profile of the System II and CO lasers measured at the burner front and back side with the knife-edging technique. Both laser systems show an excellent alignment with a horizontal tilt of less than 40 arc seconds. The diameter of the System I is diverging from a half width diameter of 180 μm to 420 μm. Hence, for this system, the spatial resolution is slightly lower than the desired value of 0.2 mm, as only ≈ 40 % of the laser intensity lied in the ±0.1 mm range around each measurement point. However, as can be seen later, the variations in the CO profiles appear on significantly larger spatial scales, which makes a lower spatial resolution acceptable. The lasers of System II were fairly well collimated with a beam diameter decreasing from 320 μm to 270 μm, which is around the desired resolution. The achieved tilts and beam diameters of both laser systems in vertical direction were similar to the horizontal direction.

Similar to the investigations by the NIST [241,242], profiles of the thermochemical state were measured at heights of 7 mm, 9 mm and 11 mm above the burner surface with a horizontal spacing of 0.2 mm. At each position, 10 000 spectra for each spectrometer were recorded. After averaging over 500 spectra, the Voigt profiles were fitted onto the spectra, yielding 20 measurement values of temperature and species concentrations at each position.

4.1.3 Post-Processing

Even for purging of the burner slots and the ambient of the grin-lens with water-vapor free gases of high purity, measurements indicate a minor water vapor absorption ($OD_e \sim 10^{-2}$) remaining in the optical path. This absorption is presumably due to ambient water vapor in the spacing between the multimode fiber and the photodiode as well as inside the laser itself [247]. In order to correct for parasitic absorption by ambient water and to effectively suppress its influence on the temperature and concentration determination, the parasitic absorption is measured by purging the burner and optical setup with dry N_2 (H_2O content < 10 ppm) prior to and after the measurements. In the post-processing, the parasitic water vapor concentration is linearly interpolated between both measurements and subtracted from the absorption spectrum of the flame measurements.

4.2 Evaluation and Results: Thermochemical State

Before comparing the thermochemistry measurements to the NIST reference data, the measurement errors and uncertainties inherent to the measurement system will be discussed. Afterwards, the developed measurement system is validated against a well-known air-blown flame. At last, the laminar, non-premixed combustion of CH_4 under air is compared to the combustion under oxy-fuel combustion atmosphere.

4.2.1 Error Analysis

As mentioned in sec. 3.4, a combined measurement error of precision and accuracy is used for simplicity, even if this may overestimate the real error. As discussed in sec. 3.4 and 3.5, the errors depend on the fit uncertainty amongst other effects. For a quantification of these, the measured spectra and fits of the air-blown operation at exemplary lateral positions 9 mm above the burner are shown in fig. 4.4.

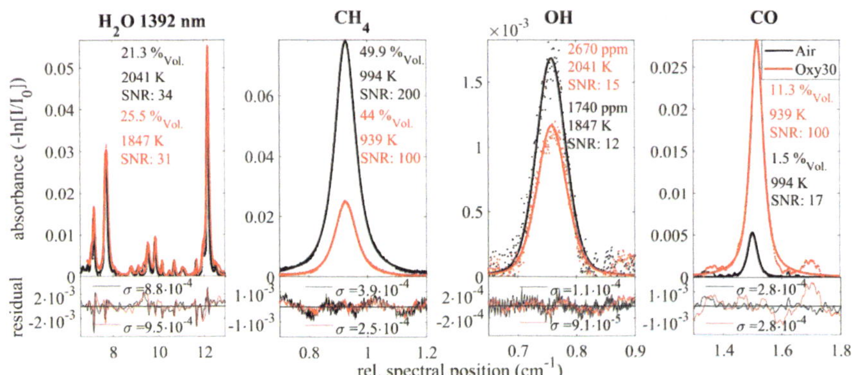

Fig. 4.4: Measured and fitted spectra of the H_2O, CH_4 and OH and CO spectrometers under air-blown and oxy-fuel combustion atmospheres (dots: experimental data, solid lines: fit)

To investigate the most complex environments for each spectrometer, the H_2O spectrum is obtained within the reaction zone at x=6.4 mm and yields a temperature of $\approx$ 2040 K. The absorption in this region is low due to a low line strength and gas density which is caused by the high temperatures. Additionally, large deviations between the measured and fitted spectral shape are evident, presumably originating from inaccuracies in the databases with respect to the line intensity, position and broadening of minor absorption lines at high temperatures. Hence, the signal to noise ratios (SNR) are low compared to the flame flanks and lead to high detection limits and measurement errors (~6 %$_{rel.}$ temperature uncertainty).

The SNRs in the colder flame flanks outside the reaction zone reach values of 100 and the temperature uncertainty decreases to 2 %$_{rel.}$.

The spectrum of CH_4 in fig. 4.4 is measured at x=2 mm, as the temperature was not measurable with the 1392 nm H_2O spectrometer in the burner center (x=0 mm) with peak CH_4 concentrations. This effect will be described later. The residual in this spectrum mainly consists of noise and what seems to be baseline nonlinearities. As the interference of other spectral lines within this spectrometer is negligible, high SNRs and consequently low measurement errors are achieved. The improved line strength uncertainty by Gharavi et al. [208] allowed to significantly reduce the errors compared to those derived in [E5] without a significant change of the measured values.

Similar to H_2O, the spectra of OH and CO in fig. 4.4 are measured in the reaction zone (x=6.4 mm) where high-temperature uncertainties and low absorbance of the species increase the respective concentration uncertainties. The SNRs of OH are in comparison lower than those for H_2O. While the residual of the corresponding spectrometer is dominated by electrical noise due to the low absorbance of the species, the CO residual shows a weak background of multiple high-temperature methane absorption lines. These lines limit the quality of the fit and thus the sensitivity, as the influence of these lines on the background cannot be accounted for sufficiently well. The uncertainty in the OH measurement is furthermore increased by the high absorption line strength uncertainties of 20 %. The absorption of CO in an oxy-fuel operated flame is significantly stronger than under air operation.

Table 4.3: Typical detection limits and measurement errors of the spectrometers used at the WHP.

System	Central wavelength target species	#		SNR	Detection limit χ_{lim}		Uncertainty	
					absolute	per length	absolute	Relative
I	2314 nm CO	1	χ_{CO}	17	800 ppm	32 ppm·m	±1 030 ppm	±11.4 %$_{rel.}$
II	1653 nm CH_4	5	χ_{CH4}	200	0.28 %$_{Vol.}$	115 ppm·m	±1.9 %	±2.9 %$_{rel.}$
	1527 nm OH	7	χ_{OH}	15	180 ppm	7.2 ppm·m	±580 ppm	±22 %$_{rel.}$
	1392 nm H_2O	8	χ_{H2O}	34	0.7 %$_{Vol.}$	287 ppm·m	±1.5 %$_{Vol.}$	±7.1 %$_{rel}$
			T				±120 K	± 5.8 %$_{rel.}$

Table 4.3 summarizes selected detection limits and measurement uncertainties that were achieved with the respective spectrometers during the investigations of the laminar, non-premixed WHP setup. As the measurement error and the standard deviation of the measurements are similar, both will be given later on.

4.2.2 Air-blown combustion and validation measurements

Fig. 4.5 shows a comparison of the horizontal thermochemical profiles between the reference measurements (black lines) by the NIST [241,242] and the measurements of this work (blue dots), exemplarily at a height of 9 mm above the burner surface. The blue error bars indicate the measurement uncertainties as derived in sec. 3.4, whereas the red bars represent the standard deviations of the measurements. The latter are caused by both statistical measurement errors and fluctuations of the burner. Due to the high stability of the burner, the standard deviation of the measurements is typically lower than the measurement error, which is dominated by systematic errors.

The temperature shows two distinct peaks due to the heat release in the reaction zone of the two flame branches. In this zone, the concentration of the reactants approaches stoichiometry and the temperature is close to the adiabatic flame temperature of methane in air at this particular equivalence ratio (2223 K). The agreement between the measured temperature and the reference data is reasonably good, with a mean relative discrepancy of only 4.1 %.

The high concentration of unburned CH_4 in the flame center and simultaneously low H_2O concentration are causing a systematical error to the temperature measurement, due to a partial overlap of the H_2O

spectrum with the CH_4 spectrum. Here, the water vapor lines cannot be resolved sufficiently well, as numerous CH_4 absorption lines with only poorly known foreign broadening and line strengths are relevant in the spectral region of the laser. As the knowledge of the temperature is crucial for the calculation of the species concentration, the species concentrations cannot be determined in the center of this flame either. Water vapor is produced in the reaction zone in the vicinity of the temperature peaks and is diffusing laterally in both the fuel lean and rich zone. The H_2O measurements show very similar trends and absolute values as the reference data, in particular in the lean branches of the flame. In the reaction zone at peak concentrations, the measurements of this work show slightly higher water vapor concentrations and fluctuations. This effect may arise from remaining flickering of the flame sheets during the measurement, which can be seen by the increased standard deviation of the measured values in the vicinity of the concentration peaks as well as from the high temperature measurement error. Deviations from the reference outside the error margins of the measurements in this work may also originate from an error in the mass flow rate, which is in the order of 1 %. It has to be noted that the error of the reference data is unknown.

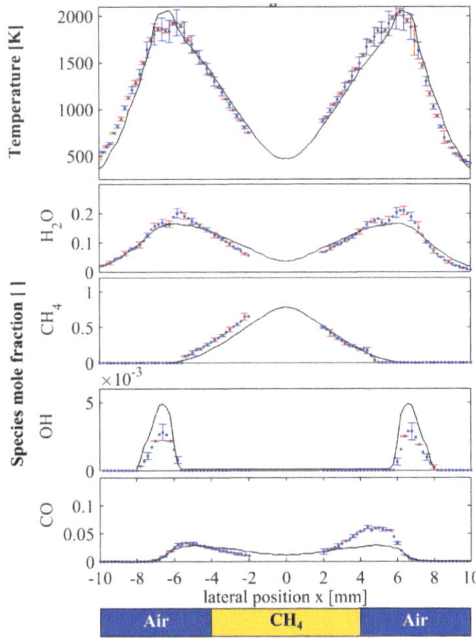

Fig. 4.5: Comparison between NIST-measurements (solid line) and measurements of this work (blue dots).

The reactant methane is ejected from the central slot in a top-hat profile. The sharp edges of this profile are soon smoothed due to diffusion. Furthermore, the CH_4 in the mixing layer approaches stoichiometric conditions in the reaction zone, where it is consumed by combustion reactions. Fig. 4.5 shows the good agreement between the laser-based measurements and the flame reference data from the NIST (mean absolute deviation: 3 %$_{Vol.}$). The CH_4 concentration is slightly increased on one branch of the burner. This effect is probably due to a minor asymmetry in the flow field, caused by either a slightly increased flow rate of the oxidizer in one slot or an asymmetric draft into the exhaust hood. High OH concentrations indicate the reaction zone, as the radical OH is an important intermediate in the combustion process, but does not significantly diffuse due to a short lifetime. As can be seen in fig. 4.5, OH exists on the lean side of the flame sheets, with a peak value at x= ± 6.4 mm from the burner center. The position of the peak OH concentration and the profile agree well between the TDLAS-based measurement and the reference data performed with LIF by Smyth et al. However, the measured absolute values are slightly less than half of what is expected from the reference data. The origin of this discrepancy could not be clarified but is significantly outside the error margins of the measurement. Nevertheless, qualitative values are sufficient to determinate the flame front by using the TDLAS system.

CO generated in the reaction zone is diffusing to both sides of the flame. As it is consumed in the lean region through reactions with oxygen, the concentration is increasing from the lean to the rich side. In one of the flame sheets, the TDLAS-measurements greatly matches the reference data. However, the opposite side reveals a significantly increased CO concentration which exceeds the measurement error. Therefore, the CO formation is either significantly increased on this flame branch, probably due to an oxygen deficiency caused by the same change in the flow field that is observed for the CH_4 measurements, or the change in the flow field transports a fraction of the reaction gases into the purging flow at the flame ends, increasing the measured CO concentration through an increase of the absorption length. The latter effect seems rather likely, as the H_2O concentration in this flame branch is also increased.

It can be concluded that the four-channel TDLAS measurement system is capable of resolving the reaction zone and measuring the gas temperature as well as the gas concentrations of CO, CH_4 and H_2O in-situ with low uncertainties.

4.2.3 Oxy-fuel combustion

Due to the decreased laminar flame speed and diffusitivity of the selected oxy-fuel combustion atmosphere, the slot exit velocities have to be reduced for a stable oxy-fuel operating condition. These effects have an impact on the flame structure, as can be seen in fig. 4.6. Here, the OH concentrations are shown at different heights for one half-side of the burner. For both combustion atmospheres, the reaction zone is moving slightly outwards with increasing height due to thermal expansion of methane in the central slot. The reduced O_2 diffusivity in the oxy-fuel combustion atmosphere lowers the reaction rate and thus the concentration of OH compared to air-blown combustion. This finding is in good agreement with the idea by Shaddix and Molina [40] of oxy-fuel reducing the local radical pool.

Additionally, the flame of oxy-fuel combustion is closer to the burner center, as the peak OH concentrations are reached at x=5.6 mm for oxy-fuel combustion and at x= 6.4 mm for air firing. This effect is probably due to the reduced strain rate between the methane and the oxidizer flow for the oxy-fuel combustion. The flame thickness, however, is similar for both cases (1.3 mm FWHM).

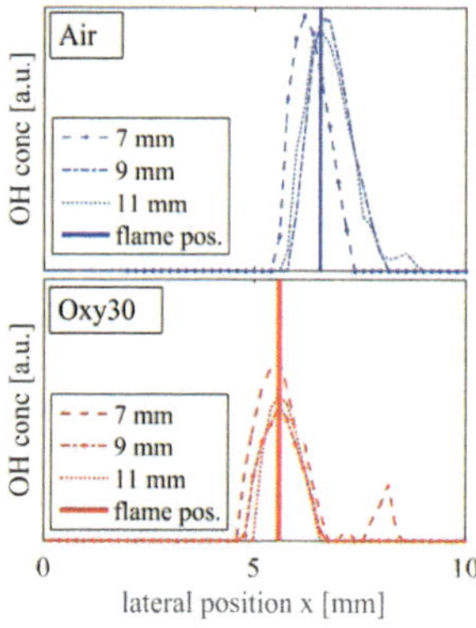

Fig. 4.6: Flame structure and OH peak position under operation with Oxy30 and air combustion atmosphere.

The thermochemical quantities of the Oxy30-flame are shown in fig. 4.7 for 7 mm, 9 mm and 11 mm above the burner surface, respectively. The gas temperatures are similar for all heights, but slightly increase with increasing distance to the burner surface. This effect is probably due to a preheating of the reactants at lower distances and therefore increased adiabatic flame temperature. From the fact that the diffusive heat transfer in Oxy30 flames is lower than in air and the mentioned effect not being observed during air-blown operation, the preheating can be assumed to originate from heat transfer through radiation of the CO_2 molecules in the reaction zone. The peak temperature and therefore the reaction zone are in good agreement with the peak OH concentration at x = 5.6 mm.

For Oxy30, the decrease of the adiabatic flame temperature by an increased molar heat capacity compared to air combustion slightly exceeds the temperature increase by a higher oxygen concentration. Kutne et al. [18] reported the stoichiometric flame temperature of CH_4/Oxy30 under adiabatic conditions to be 2160 K and thus slightly less than for CH_4/air combustion. However, the experimental peak temperatures in oxy-fuel combustion (1860 K) are $\approx$250 K lower than within air combustion at the same height and 300 K lower than the stoichiometric flame temperature of CH_4 in Oxy30. The reduction in temperature is probably due to the aforementioned increased heat transfer from the reaction zone into the reactants through radiation. The same effect has been observed by Kutne et al. with very similar temperatures.

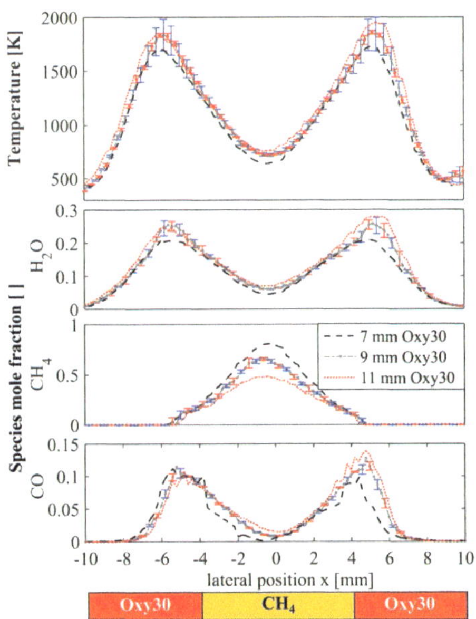

Fig. 4.7: Measurement of thermochemical quantities during Oxy30 operation in different heights above the burner.

The H_2O measurements show a significant increase in the absolute peak water vapor concentration compared to the air-blown operation. This effect can be explained by the modified composition of Oxy30 compared to air: in a stoichiometric mixture of Oxy30, the concentration of O_2 is increased. Hence, for a stoichiometric mixture, a higher CH_4 concentration is required and as such, after combustion, the H_2O concentration is increased. An equilibrium calculation of a stoichiometric Oxy30/methane mixture (fig. 2.5) predicts a H_2O concentration of 29.2 %$_{Vol.}$ in the exhaust gas. This value slightly exceeds the measured values of $\approx$26 %$_{Vol.}$ within the reaction zone, probably due to diffusion effects in the 2-dimensional setup. A similar approximation for air yields a lower water vapor concentration of approximately 18.3 %$_{Vol.}$, which is in good agreement with the experimental findings. The peak H_2O concentration for a given burner distance slightly increases downstream from the lowest to the mid position where it reaches a maximum. Unlike non-premixed air combustion, the oxy-fuel flame allows to measure the gas temperature in the central position between the flame sheets, as the H_2O concentration, and as such the absorbance, are higher. Additionally, the methane absorption is smaller due to lower concentrations. Both effects cause the impact of methane on the temperature measurement to be negligible, which allows a determination of species concentrations at the burner center.

The methane concentration is strongly decreasing with increasing distance to the burner, as the combustion products CO_2 and H_2O diffuse from the reaction zone into the fuel-rich area and dilute the methane stream. In return, CH_4 diffuses into the reaction zone where it is consumed. Compared to air, the reaction progress seems to be enhanced, as the CH_4 concentrations at same heights are lower. This effect is probably due to either an increased conversion rate caused by the increased O_2 concentration in the combustion atmosphere or the lower exit velocities of the gas streams.

Similarly to the air blown combustion, the concentration of CO is higher on the rich side of the reaction zone, as it is not oxidized to CO_2 as on the lean side. However, the concentration is significantly increased during oxy-fuel combustion, with peak values of approximately 11 %$_{Vol.}$ compared to 3 %$_{Vol.}$ with air. A comparison of these in-flame CO concentrations to equilibrium calculations, which represent the flue gas situation where CO is mostly oxidized, is not meaningful. A non-premixed calculation of the situation, however, exceeds the scope of this work. The origin of the high CO concentrations lies in the higher concentrations of CO_2 and H_2O within the oxy-fuel reaction zone: In a coarse simplification for a principle understanding, the oxidation of methane can be described by the reduced four-step reaction model by Peters [248]. This model is advantageous, as only a few reactions are considered compared to a large number of elementary reactions in detailed chemistry calculations. The reaction steps are:

$$CH_4 + 2H + H_2O \rightleftharpoons CO + 4H_2 \qquad (4.I)$$
$$CO_2 + H_2 \rightleftharpoons CO + H_2O \qquad (4.II)$$
$$2H + M \rightleftharpoons H_2 + M \qquad (4.III)$$
$$O_2 + 3\,H_2 \rightleftharpoons 2H + H_2O \qquad (4.IV)$$

Reactions (4.I) and (4.II) are the dominant reasons for an increasing CO concentration for the combustion of CH_4 under an oxy-fuel combustion atmosphere. Through the increased concentration of H_2O, originating from the increased O_2 in the oxidizer as described above, the equilibrium leads to an increased CO concentration on the product side of (4.I). Due to the significantly increased CO_2 content, reaction (4.II), which is the water-gas shift reaction, leads to a further increase in the CO concentration. These reactions are dominant in the reaction zones where all reactants are available. In the center, due to the incomplete reaction progress, the CO concentration is low and originates from diffusion only. The high concentrations of CO pollutants in non-premixed oxy-fuel combustion systems play an important role in the design of oxy-fuel burners. As the non-premixed atmosphere of volatiles around solid fuel particles burns in a similar manner to this generic setup, measures have to be taken to reduce the CO concentrations. This can be either through long residence times in the combustion chamber, which allow a higher reaction progress, through lean combustion or through an oxygen staging in the flue gas, which allows for a post-oxidation.

4.3 Conclusions

A laminar, non-premixed two-dimensional methane flame under air and oxy-fuel atmosphere was studied using multiple of the spectrometers developed in chapter 3. Reference data from the literature of the air-blown operation of the burner was used for a validation of the measurement system and found to be in good agreement. A new, stable oxy-fuel operating condition of the burner within an Oxy30 combustion atmosphere was defined and investigated through a comparison of the two flames. Even though the stoichiometric flame temperatures are comparable, a decrease in peak flame temperature was observed due to radiation losses. The flame thickness was similar among the two flames, but the oxy-fuel flame moved towards the burner center and the concentration of OH radicals decreased. The water vapor concentration in the reaction zone was observed to be increased by an oxy-fuel combustion atmosphere due to the increased oxygen content. Similarly, the concentration of CO significantly increased due to an increased carbon availability, which was explained through a simplified set of chemical reactions that lead to a significantly altered chemical composition during oxy-fuel combustion. The measured thermochemical quantities are compared to equilibrium calculations of a stoichiometric mixture. The equilibrium H_2O concentrations are found to match the measured H_2O peak concentrations. The peak CO concentrations of this non-premixed configuration are very similar to the peak CO concentrations within a calculated 1-D premixed flame. Nevertheless, the measured peak temperatures are lower than the adiabatic flame temperature due to radiative heat losses. The measured data can be utilized for future validations of numerical oxy-flame combustion models and simulations.

5 Turbulent Oxy-Gas Combustion: The Oxy-Fuel Burner

Compared to conventional air-blown combustion (section 2.1), a multitude of effects are altered or intensified in oxy-fuel combustion, but there is a significant lack of understanding of these mutually-dependent effects. Comparisons of the results of deducted models with quantitative experimental findings as well as the identification of relevant parameters for experimental measurements through simulations are indispensable to better understand such effects. A holistic understanding as well as consecutive model descriptions and simulations of relevant parameters is critical in the creation of predictive engineering tools, which are essential for the design of future oxy-fuel gas turbines or coal power plants.

The research community, in particular the collaborative research center/Transregio 129 "Oxy-Flame", follows a stepwise approach by isolating single effects in generic setups and increasing the complexity of the burners and fuels to close-to-application combustion systems. In the intermediate range of complexity and thermal power (10–100 kW_{th}), detailed and comprehensive experimental and numerical investigations of the oxy-gas and oxy-coal combustion are still sparse.

To close this gap, two complementary oxy-fuel combustion systems are developed and investigated at the TU Darmstadt and the RWTH Aachen: The *Oxy-Fuel Burner* and the WSA-*Technikum*. Both are turbulent oxy-fuel burners, which are down-firing into a combustion chamber. The first is intended to fire oxy-gas or gas-assisted oxy-coal flames, whereas the latter is primarily burning pure oxy-coal flames. Both are currently under comprehensive investigations of the flow, chemistry and thermodynamics for mutually enabling comprehensive insights. In this chapter, the oxy-gas combustion of the Darmstadt Oxy-Fuel Burner will be studied prior to gas-assisted oxy-coal combustion in the subsequent chapter.

Measurements of the thermochemical state of the flue gas of this burner allow an analysis and comparison to simulations of the reaction progress and heat loss through diffusion or radiation. Both of these effects are affected by the combustor flame position and flow field, which have already been studied by Becker et al. [23]. Local measurements of the flow quantities (e.g. velocity, shear rate and fluctuations) using PIV or LDA are already established in combustion research. However, the global result of the flow field, the residence time distribution (RTD), has rarely been studied, because only a few, complex measurement techniques allow to determine this quantity. The RTD is a measure of how much time fluid elements spend within the combustor and is highly relevant for the design of the latter, as it affects the reaction progress or burnout of coal particles. It also allows for the identification of large flow features, such as recirculation zones or plug flows, even though without delivering information on their location. Moreover, chemical reactor networks (CRN), which simulate the flow features based on an input-output analysis, can be optimized with respect to knowledge of the RTD and the thermochemical composition of the flue gas [143,144]. Compared to computational fluid dynamics, CRNs allow a higher detail on the chemistry modeling. These networks allow conclusions regarding the flow patterns in combustors and possible optimizations [249]. An optimization of the CRN with respect to experimental data can even allow for an interpolation or extrapolation to unknown operating conditions.

In contrast to the local quantities, the RTD is an integration of the velocities along the fluid element path. Hence, in situations where the experimentally measured and simulated local flow field seem to be in a good agreement, the RTD offers a more precise measure as minor differences in the local velocity add up and show large deviations in the RTD between simulations and experiments. Nevertheless, the RTD does not allow to study the origin of the deviations.

The need for global measurements of the flow field and the thermochemical compositions of the same turbulent oxy-gas combustion system define the goals of this chapter:

- to develop a measurement technique which allows for reliable measurements of the residence time distribution under the harsh environments of oxy-fuel combustors and to analyze the potential and limitations of this technique,

- to apply the measurement technique to the Oxy-Fuel Burner and measure the RTD under non-reacting and reacting conditions of air-blown and oxy-fuel gas combustion. This stepwise increase of complexity allows for analysis of changes in the RTD and identification of flow features,

- to measure the thermochemical composition of the gas flame under air, Oxy25 and Oxy30 combustion atmospheres and to compare the results to equilibrium calculations. This allows an identification of effects which are not described by this approach.

A simultaneous simulation of the combustor flow field and thermochemical state by CRNs, based on the measurements or computational fluid dynamics (CFD), exceeds the scope of this work and will therefore not be included.

The first section explains the experimental setup of the thermochemical and RTD measurements. Afterwards, the techniques for measuring the RTD are evaluated and the results of the measurements are presented and analyzed. In the third section, the thermochemical state is investigated based on TDLAS-measurements and compared to equilibrium calculations. Parts of this chapter have recently been published by the author in [E1] and [E6-E10].

5.1 Experimental Setup

First, the design and chosen operating conditions of the Oxy-Fuel combustor will be described. Two strategies and associated setups of RTD measurements will be presented afterwards, followed by necessary adaptions of the general optical setup, which was presented in subsection 3.3.

5.1.1 Oxy-Fuel Burner Setup

Even though the Oxy-Fuel Burner is primarily designed to fire gas-assisted oxy-coal flames, it is also able to sustain the same flames in air or fire pure gas flames in air or oxy-fuel combustion atmospheres. Firing pure methane flames, the burner is operated with thermal powers up to 20 kW$_{th}$ and aims to close the gap between unconfined laboratory-scale burners [37,38] on the one hand and confined close-to-application combustion systems [31,32] or even power plants [115] on the other. The burner itself (fig. 5.1a) is designed to mimic the features of a coal firing burner [250] and is down-firing into the combustion chamber of 420 x 420 x 600 mm^3 volume (fig. 5.1b). Through the aid of guidance plates, the corners of the combustion chamber are rounded. Wedged, fused silica windows in the walls allow spacious optical access.

Similar to industrial coal burners, the gas issues from the burner through a partially premixed (fuel and oxidizer), unswirled primary flow and a swirled secondary oxidizer flow. While the ratio between the primary and secondary flow allows for a variation of the local, near burner equivalence ratio and swirl number, a tertiary flow close to the windows allows for a staging and a variation of the global equivalence ratio. The primary and secondary flows issue from two concentric annular orifices atop of a quarl in which the two flows mix and a flame can be stabilized. This transparent quarl is of particular relevance for gas-assisted oxy-coal combustion, as the coal particles from the primary flow are contained within the high temperature environment of the flame for a longer time and therefore show a higher burnout rate, reaction progress and flame stability.

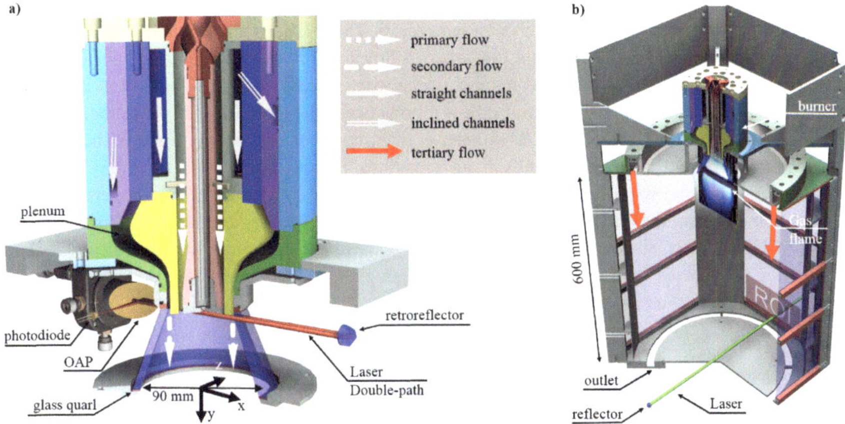

Fig. 5.1: Sectional view of the Oxy-Fuel Burner with combustion chamber: a) Burner with flows, optically accessible quarl, coordinate system and single-sided doublepass for RTD-measurements; b) whole combustor with chamber, gas flame and doublepass for RTD and thermochemical measurements.

The flue gas measurements are performed in the lower quarter of the combustion chamber where the flue gas exits through an annular outlet orifice at the bottom. The pressure within the combustor is slightly higher than the ambient pressure to avoid leakage of ambient air into the combustion chamber through holes.

Becker et al. [23] measured the flow field and flame stabilization within the quarl of this burner under reacting and non-reacting conditions using stereoscopic particle image velocimetry and LIF of the OH molecule (OH-LIF). Using the same combustor, Doost et al. [E7] simulated the non-reacting flow field and residence time.

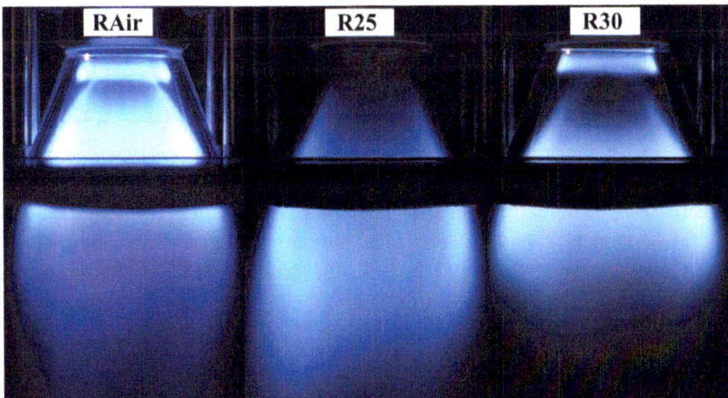

Fig. 5.2: Photography of the methane flames (20s exposure time) [23].

While this burner allows to stabilize the aforementioned pure gas flames, it requires a thermal stabilization of the flame for oxy-coal combustion because a large proportion of the heat necessary for the ignition of solid fuel particles is lost through the walls due to radiative and diffusive heat transfer. Hence, solid fuel flames are stabilized by assisting gas flames. In this chapter, pure methane flames under the three combustion atmospheres are investigated. In the following chapter, these flames serve as stabilizing flames for the gas-assisted oxy-coal combustion.

The three different operating conditions of pure methane flames are identical to the previous work [23]: the combustion of methane under air, Oxy25 and Oxy30. These flames are termed RAir, R25 and R30 (fig. 5.2). Under non-reacting conditions, the same gas flows are referred to as NRAir, NR25 and NR30. The thermal power as well as the swirl number of the reacting cases are kept constant at 20 kW_{th} and approximately 0.47 (table 5.1). The impulse flow ratio between the three flows in the burner is constant for all three operating conditions. The Re-number of the primary and secondary flows in the orifices all lie in the range of 2500 and 3100 and thus in the transitional regime. The premixed primary flow is fuel rich, but due to mixing with the secondary flow in the quarl and in particular with the tertiary flow in the post-flame region, the equivalence ratio (Φ) is globally lean. This is necessary to allow a co-firing of coal at higher thermal power in the next chapter without the formation of a globally rich flame. However, it comes with the drawback of artificially lean global equivalence ratios and a therefore high oxygen excess in the flue gas.

Table 5.1: Operating conditions under investigation: Equivalence ratio, oxidizer, power, Re-number and volume flows under standard conditions [23].

	Non-reacting			Reacting		
	NRAir	NR25	NR30	RAir	R25	R30
Oxidizer	Air	Oxy25	Oxy30	Air	Oxy25	Oxy30
Fractions Air/O_2/CO_2 [%vol.]	100/0/0	0/25/75	0/30/70	100/0/0	0/25/75	0/30/70
Primary equivalence ratio (Φ_I)	-	-	-	1.6	2.4	1.97
Quarl equivalence ratio (Φ_{I+II})	-	-	-	0.64	0.91	0.75
Global equivalence ratio (Φ)	-	-	-	0.19	0.27	0.22
Re-number primary flow	2765	2649	2599	2839	2650	2609
Re-number secondary flow	3080	2998	2932	3080	2988	2932
Thermal power		-			20 kW_{th}	
Swirl number		0.47			0.47	
Volume Flows (m^3/h)						
I CH_4	-	-	-	2.01	2.01	2.01
I Oxidizer	13.55	8.05	8.16	12.00	6.68	6.78
II Straight Oxidizer	5.97	3.75	3.76	5.97	3.75	3.76
II Inclined Oxidizer	12.02	7.18	7.27	12.02	7.18	7.27
III Oxidizer (m3/h)	69.95	42.16	42.60	69.95	42.16	42.60

For a stepwise increase of complexity in terms of necessary measurement techniques and flow structure, the RTDs of the flows of each flame are investigated not only under reacting conditions, but also under non-reacting conditions first (NRAir, NR25 and NR30). Here, the methane flows are replaced by the respective oxidizers of the same momentum flux.

5.1.2 RTD Measurement Setup

In general, measurements of the RTD are performed by injecting a non-reactive tracer into one or multiple inlet-flows and probing the outlet concentration as a system response. Two canonical strategies exist to measure the RTD: the pulse and the step injection response (section 2.3). Both are applied to the Oxy-Fuel Burner for an evaluation of their respective limitations and potentials. Only few molecules are suitable as tracers, of which hydrogen chloride (HCl) has been chosen for the following reasons:

- *Stability in combustion processes:* The bonding energy of HCl (431 kJ/mol) significantly exceeds that of most hydrocarbons and is approximately half of the N_2 value. In addition, only few stable molecules can be created out of chloride-hydrocarbon reactions. These proberties promise a low reactivity in combustion environments,

- *High detectability:* Section 3.1 shows that the line strength of HCl exceeds those of most other species, which allows a low detection limit,

- *Density similar to oxy-fuel combustion atmosphere:* The density of HCl (1.63 kg/m^3) is very similar to oxy-fuel (Oxy25: 1.65 kg/m^3; Oxy30: 1.59 kg/m^3). This prevents a separation of the tracer from the combustion atmosphere through gravitational or diffusional effects, and the injection does not have an influence on the momentum flow.

However, safety precautions must be made during operation with HCl due to its toxicity and corrosiveness. In [E1], it is shown that HCl and CH_4 tracers yield similar results in non-reacting conditions. As the latter cannot be used under reacting conditions, HCl will be used for all experiments in this work for simplicity.

The tracer is injected into the combustion chamber's primary flow 25 cm upstream of the burner by using a quickly reacting solenoid valve (opening time ~37 ms). As the tracer mass flow rate needs to be negligible compared to other flows, an 700 μm aperture restricts the HCl concentrations to less than 1.6 %$_{Vol.}$ of the primary flow rate (<0.21 %$_{Vol.}$ of the overall flow rate). During step injection measurements, the valve is opened and closed for 20 s each in a series of 50 injections. In the pulse measurements, the valve is open for 200 ms and then closed for 20 s, with 50 injections for each operating condition as well. The concentrations are simultaneously measured at the inlet and outlet using the 1742 nm HCl spectrometer and recorded together with the valve triggers. The laser repetition frequency was 2 kHz. Under non-reacting conditions, ten absorption spectra were averaged before spectral fitting, and five under reacting conditions, leading to effective measurement rates of 200 Hz and 400 Hz, respectively.

5.1.3 Optical Setup

The tracer inlet concentration is measured 1 mm downstream the annular orifices of the primary and secondary flow atop the optically accessible quarl (fig. 5.1a), where a single-sided double-pass setup is set up. This double-pass arrangement is necessary as a short absorption path (four passes across the 3 mm orifice, L_{abs}=12 mm) and severe beam-steering due to the turbulent flame in the quarl are expected. The conical surfaces of the quarl inherently prevent self-interfering fringes.

The outlet concentration is measured through the combustion chamber center (x=0 mm) 35 mm upstream the outlet plate (y=565 mm). During step injections, the expected concentrations are similar to those simulated in subsection 3.1.7, which suggest a path length in the order of one meter. Hence, a single-sided double-pass with an absorption length of L=82 cm is utilized (fig. 5.3, red lasers), which consist of a hollow OAP with a broadband collimator located inside and a fused silica retroreflector with AR-coating on the opposing side. For the pulse injection, significantly lower outlet gas concentrations are expected due to a shorter injection duration and a therefore reduced accumulation of tracer molecules. Hence, a single-sided multi-pass arrangement with a total of eight transitions and L=3.36 m is established through the chamber center using three adjustable planar gold mirrors and a hollow OAP (fig. 5.3, blue lasers). The light of the 1742 nm HCl spectrometer is separated into two fibers using a 50:50 fiber splitter and guided to the inlet and outlet measurement positions.

The thermochemical state of the flue gas is measured using all spectrometers listed in table 3.1, exception for the 1742 nm HCl laser as HCl is not expected in the combustion of methane. During these measurements, the combustor is moved with respect to the optical setups of System I and II to perform line-of-sight measurements at different axial and radial positions. The measured region axially covers the most downstream quarter of the burner (465 mm$\leq$ y $\leq$ 565 mm) and radially spans along the central half (-105 mm $\leq$ x $\leq$ 105 mm). Closer to the flame, the flow is less mixed, which prevents a line-of-sight measurement. The guidance plates prevent measurements further radially outwards due to laser-beam blocking. For these systems, the same doublepass arrangement as during the step-RTD measurements is used sequentially (fig. 5.3, red lasers).

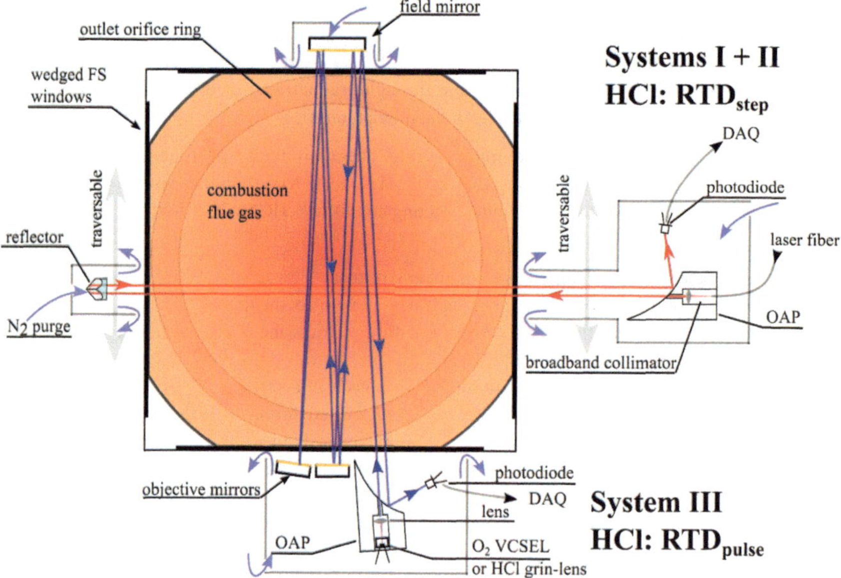

Fig. 5.3: Top view of the combustion chamber and the optical setups for RTD and thermochemical measurements (not in scale).

System III, which measures concentrations of O_2, utilizes the identical beam arrangement as for the pulse-RTD experiments, only with the laser directly placed within the hollow OAP. Due to the sensitivity of the multi-pass setup to misalignments, the burner is not moved with respect to this optical setup. All chamber windows are vertically wedged to prevent fringes.

5.2 Evaluation and Results: Residence Time Distribution

At first, the different approaches to measure the residence time distribution are compared. Afterwards, one of them is selected for measurements at the Oxy-Fuel Burner under non-reaction and reacting conditions. The results are discussed and compared to simple models of the reactor.

5.2.1 Comparison of Techniques

Fig. 5.4a shows the input signal of a positive (up) and a negative (down) injection step, exemplarily for the NR30 operating condition. The signals are averaged over 50 injections each. At t=0, the valve is opened or closed, respectively. The signals of negative step injections are always subtracted from one to calculate the CDF: $F(t)=1-W(t)$ within outwash experiments. Both steps induce a steep change to the input concentration, which quickly approaches a steady state. The positive step shows minor fluctuations at the beginning, which are due to pressure oscillations in the supply tube from the gas bottle to the valve. The negative step does not exhibit this feature, as oscillations after valve closing do not affect the injection.

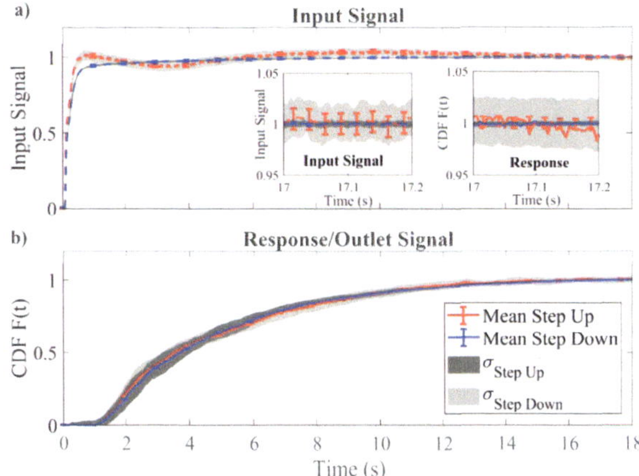

Fig. 5.4: a) Input signals and standard deviation of a positive and negative step injection.
b) Signal responses (CDF) of the injections, standard deviations and errors.

The onset of the CDF (fig. 5.4b) is delayed to the input signal by approximately 1.5 s and is followed by a smooth rise of the tracer concentration. Both injection strategies exhibit very similar mean cumulative distribution functions, with a correlation coefficient of 0.9992 between the two signals. Mainly, the CDF gained from the positive step shows a minor signal reduction approximately 2 s after tracer onset, which is probably due to the described fluctuations in the injection. Nevertheless, the mean responses of the two strategies after 50 injections are practically indistinguishable.

However, the phase-averaged standard deviation (STD) of the CDF of the negative input is smaller, which can be seen in the inserts in fig. 5.4a: the two inserts show the excitation signals and the CDFs by the two step types for a single injection after reaching steady state. In addition, the standard deviations and errors of 50 injections are displayed. Assuming constant conditions in the reactor, the uncertainties in line intensity, temperature and pressure do not account for errors in the RTD or CDF. Therefore, only line area errors by deviations from the spectra and the line model, expressed by the SNR, are considered.

The standard deviation and measurement error of the negative step are significantly smaller than for the positive and therefore not visible in the insert of fig. 5.4a. This general behavior has also been observed for turbulent systems by Nauman et al. [139] and in previous experiments [E1]. The effect was attributed to a reduced error of the final value, but lacked a detailed explanation.

Fig. 5.5 shows the phase-averaged standard deviations of the two injection strategies over the injection time. For the positive step, the absolute STD is initially small (region (I)), as the tracer concentration is zero and variations therefore originate in measurement noise only. In turbulent systems, any two injections are statistically independent from each other if the time between the two injections exceeds the integral time scale, which is the case for the measurements presented here.[30] Hence, the temporal evolution of the CDF is different for each injection, because the tracer is exposed to different flow conditions. In region (II), where the shape of the CDF is significantly affected by these variations in the flow, the standard deviation is high and mainly depends on the turbulent intensity. For a positive step, the STD remains high when approaching the steady state (region (III)), as the tracer concentration is

---

[30] The integral time scales were deducted from [E7].

fluctuating due to turbulent eddies. These fluctuations make a determination of the steady state and thus a subsequent normalization of the CDF challenging and prone to errors.

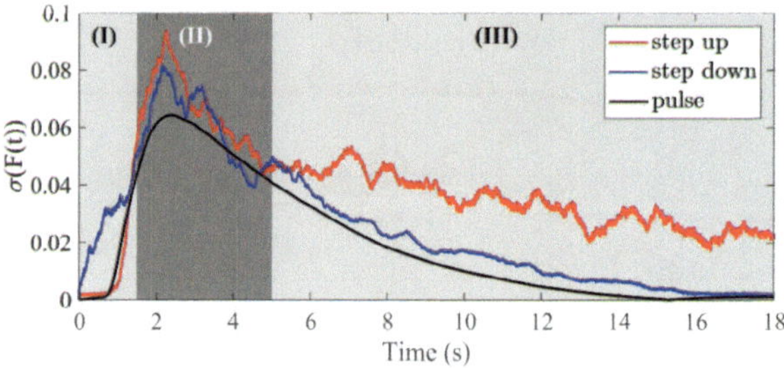

Fig. 5.5: Standard deviations of the CDF for different injection types (NR30).

For a negative step, the behavior is reversed. Here, the final value (region (III)) is close to zero and therefore only depends on measurement noise and not on turbulence. This leads to a small STD and causes washout experiments in turbulent systems to appear smoother and to seemingly have a lower final value variance. In return, however, their STD is high for the initial value whose value is fluctuating due to turbulence (I). In similarity to the final value in a positive step injection, this makes a determination of the initial value and thus a subsequent normalization prone to errors.

In region (II), where the STD is dominated by flow variations and consecutive variations in the RTD, both techniques show similar STDs. Hence, they are both equally good in determining the CDF. The error of the mean CDF $\varepsilon_{F(t)}$ in turbulent systems can be estimated through [251]:

$$\varepsilon_{F(t)} = \frac{\sigma(F(t))}{|\max(F(t)) - \min(F(t))|} \cdot \frac{1}{\sqrt{N_{inj.}}} \qquad \text{(Eqn. 50)}$$

with N_{inj} the number of injections. As the denominator of the first term is approximately one for the CDF, the first term $\sigma(F(t))$ which is the phase-averaged standard deviation of the measurements due to turbulence, is less than 0.1 for the NR30 operating condition presented in fig. 5.5. Therefore, for 50 injections, the error of the mean CDF is around 1 %, independently of the injection scheme.

In theory, a CDF measured by step injections allows a determination of the RTD $E(t)$ by taking the temporal derivative $E(t) = dF(t)/dt$ (section 2.3). However, as discussed in [144] and [E1], this process significantly amplifies the measurement noise and fluctuations from the turbulence. Even with severe filtering or averaging, a meaningful determination of the RTD is not possible in turbulent systems using the CDF. However, it can be determined indirectly by modeling of the flow through multiple elementary (plug-flow or well-stirred) reactors and fitting of the analytical model to the experimental CDF [E9]. The derivative of the analytical model function is not influenced by turbulence and noise and therefore yields more precise and meaningful results. However, this modelling requires presumptions of the flow structure. The CDF can also be compared to step injection responses in CFD simulations of the flow [E7] for validation.

Pulse injections allow a direct measurement of the RTD. Fig. 5.6 shows the pulsed inlet signal and the induced response of the NR30 operating condition, averaged over 50 injections. The pulse has a duration (FWHM) of 240 ms, which is slightly longer than the valve opening time due to mechanical inertia in the valve. The inlet signal is very reproducible with a peak STD of <3 %. As the injection into the primary flow is performed 25 cm upstream of the quarl inlet and measurement position, a minor delay is evident between valve opening and tracer peak concentration. This delay depends on the operating condition, due to changes in the flow velocity, and is subtracted from the RTD values.

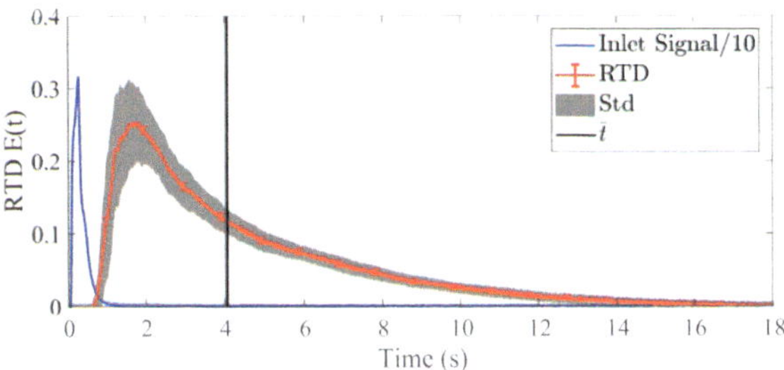

Fig. 5.6: Pulse injection and residence time distribution signal at the outlet (NR30).

Similarly to the step injections, the response is delayed by approximately 1.5 s. Afterwards, it shows a steep rise due to the majority of tracer molecules arriving at the outlet measurement position. Due to the described turbulence in the flow, the absolute STD of the RTD is high around the peak values. Afterwards, the RTD shows an exponential decay. The measurement error, estimated from the spectral fit uncertainty, is significantly lower than the turbulent fluctuations.

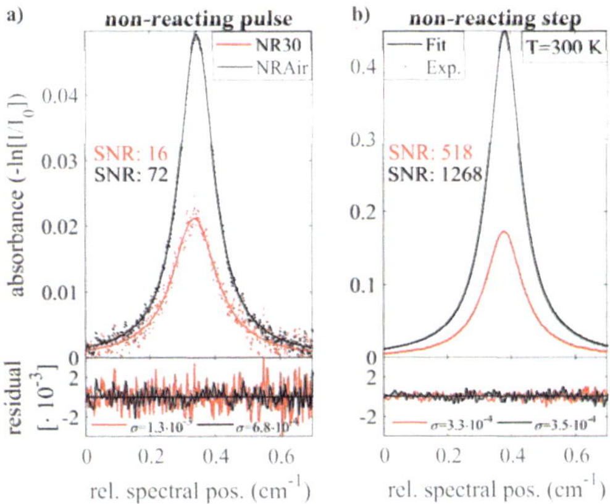

Fig. 5.7: Absorption spectra of tracer HCl under non-reacting conditions at peak concentrations: a) Pulse injection, HCl peak concentrations: ~55 ppm (NR30)/ ~100 ppm (NRAir). b) Step injection, HCl peak concentrations: ~0.15 % (NR30)/ ~0.35 % (NRAir).

Fig. 5.7 shows absorption spectra of HCl at peak absorbances of pulse (a) and step (b) excitation. For comparison, spectra of the non-reacting conditions NR30 and NRAir are shown. Even though the absorption path length of the setup used in pulse experiments is four times longer than for step experiments, the SNR is 20–35 times smaller. There are two reasons for this reduction:

1. *Reduced tracer concentrations:* Due to the accumulation of tracer molecules, the concentrations in step experiments are ~30 times higher than for pulse experiments.

2. *Increased measurement noise:* The increased absorption path length in the pulse experiment increases not only the absorption, but also the measurement noise due to amplified beam-steering by a factor of 2–4, which partially compensates the benefit of an increased absorption length.

Due to the first reason, the relative measurement errors using identical measurement techniques are typically significantly larger for pulse injections. The increased errors require counteracting by an increased experimental effort to achieve lower detection limits. As most measurement techniques do not allow a reduction of the detection limit, most measurements in literature rely on step injection to obtain the CDF of combustors [133,144,145] even though this technique impedes a direct determination of the RTD.

Fig. 5.7 additionally shows that the absorption line is collisionally-broadened by the presence of Oxy30, as observed in subsection 3.5.3. Furthermore, flow velocities are shorter, leading to a longer mean residence time in the combustor and an increased dispersion of the tracer profile during NR30 operation, which subsequently reduces the tracer concentration. The broadening in combination with a reduced tracer concentration decrease the SNR further.

As described, the errors of the CDF are amplified by transformation into the RTD. Hence, a comparison of pulse and step injections by means of the RTD is not very meaningful. Instead, the injection types are compared by transforming the RTD $E(t)$, obtained by pulse excitation, into the CDF by integration through $F(t) = \int_0^t E(\tilde{t})d\tilde{t}$. Fig. 5.8 shows the result of this transformation in comparison with a direct CDF measurement through a negative step injection. The excitation signals of both strategies are very similar, as are the CDFs. The measurement errors are higher for the pulse injections due to lower tracer concentrations, but are still significantly undercutting the STD. These observations allow to use the pulse injection measurements not only to calculate the RTD but also the CDF.

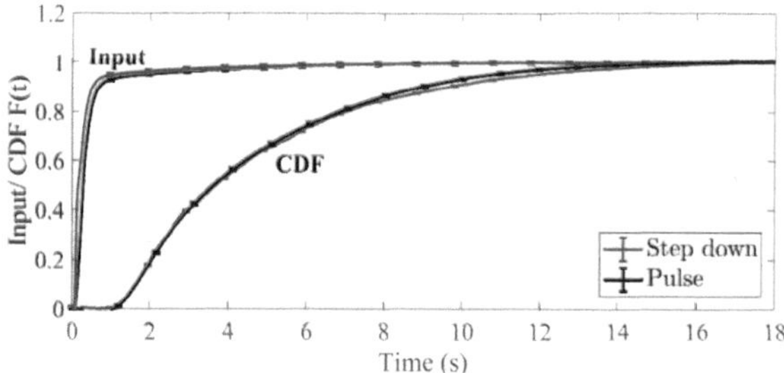

Fig. 5.8: Comparison of CFD obtained by negative step injection and pulse injection (RTD transferred into CDF) for NR30.

Similar to the mean CDF gained from pulse experiments, its STD can be obtained through integration of the RTD, which is shown in fig. 5.5 in comparison to the STDs from the positive and negative step injection. In contrast to the latter, the CDF is low in both region (I) and region (III), because the tracer concentration is zero before and long after the injection. In region (II), where the shape of the CDF is dominated by the turbulent variations of the flow, the three strategies show a similar STD.

In summary, a direct measurement of the RTD using pulse exaction/injection is preferable, as

- the CDF can be obtained by both techniques, but more reliably using pulse experiments because steady state values are not required for normalization,

- in contrast to CDF measurements using step excitation, RTDs can be measured without filtering or modeling. Hence, important quantities of the flow such as mean residence time or the higher statistical moments can be directly calculated.

- the RTD can be measured in sensitive flows or flames because the short injection duration and significantly lower concentration have less influence. As an example, the R25 operation was too sensitive for step injection measurements, as even the low concentrations of <2 %$_{Vol.}$ tracer in the primary flow visibly altered the flame shape. No changes on the flame, however, where evident during the significantly shorter pulse injection.

On the downside, measuring the RTD using pulse injections requires a more sensitive measurement system because the tracer concentrations are significantly lower. This increases the experimental effort as a multi-pass arrangement must be established and it increases the measurement errors. However, it was possible to demonstrate, that errors can be kept below the turbulence-induced STD of the RTD.

For these reasons, the results presented in the following subsections were obtained by pulse injection measurements. To achieve uncertainties of the average RTD in the order of 1 %, 50 injections were measured each.

5.2.2 Non-reacting Conditions

In this work, the measurements of [E1] were repeated for all operating conditions using pulse injections rather than step injections in combination with a subsequent modeling of the flow. The reason for this repetition is a distortion in the absolute values of the publication. This publication primarily described a measurement strategy for the mean residence time. While the strategy is still valid, the quantitative results for the given operating conditions, however, were distorted due to severe leakages in the combustion chamber which were discovered after publication. In addition, because of the effects described in the previous subsection, measurements by means of CDF and subsequent modeling are less reliable than direct measurements using pulse injections. For these reasons, the measurements were repeated using pulse injections.

From the RTDs gained by these measurements, the mean residence times $\overline{t}$ can be directly calculated using eqn. 46. Table 5.1 summarizes those of the three non-reacting conditions. The uncertainties represent the standard deviations of the parameters, which are obtained by calculating the mean residence time from individual injections.

In literature, the mean residence time is mostly estimated through the ratio of the combustion chamber volume and the gas flow rate, $V/\dot{V}$. However, as table 5.2 shows, this is a very crude assumption for complex flows, even under non-reacting conditions without expansion. As the mean residence times $\overline{t}$ of all the operating conditions are shorter than the ratio, recirculation zones can be assumed to occur in the flow, which reduce the effective volume of the combustion chamber.

Table 5.2: Mean residence time, variation and skewnesses of the investigated operating conditions together with the respective standard deviations due to turbulence in the flow.

	Non-reacting			Reacting		
	NRAir	NR25	NR30	RAir	R25	R30
$V/\dot{V}$ [s]	3.84	6.22	6.22	-		
$\overline{t}$ [s]	$3.37^{\pm 0.28}$	$4.06^{\pm 0.35}$	$4.05^{\pm 0.35}$	$1.26^{\pm 0.47}$	$2.35^{\pm 0.23}$	$2.46^{\pm 0.24}$
$\overline{t}_{PFR}$ [s]	0.81	1.01	1.01	0.43	0.88	0.81
$\overline{t}_{WSR}$ [s]	2.56	3.05	3.04	0.83	1.47	1.65
σ_{RTD} [s]	$2.74^{\pm 0.07}$	$3.28^{\pm 0.08}$	$3.25^{\pm 0.05}$	$1.16^{\pm 0.48}$	$1.83^{\pm 0.20}$	$1.96^{\pm 0.17}$
s_{RTD} [s]	$1.18^{\pm 0.05}$	$1.11^{\pm 0.03}$	$1.11^{\pm 0.02}$	$2.60^{\pm 2.39}$	$1.24^{\pm 0.31}$	$1.18^{\pm 0.14}$
Bo	0.22	0.11	0.12	1.33	0.51	0.50

The two oxy-fuel operating conditions are very similar, as their flow velocities and densities differ only little. The mean residence time of NRAir is shorter due to an increased flow velocity. The same holds true for the residence time variance σ_{RTD} (section 2.3), which is shorter for NRAir. It describes the dispersion of residence times around the mean and therefore shows that the mixing is enhanced for

NRAir, probably caused by a higher turbulence level through the increased *Re*-number in the primary and secondary flows. Due to this turbulence, each injection leads to a slightly different response. The variation of the mean residence time is in the order of 8 % for all operating conditions, which is similar to the findings in fig. 5.5. This variance is not identical to the residence time variance σ_{RTD}: the first describes how the mean residence time scatters between different injections and would be zero for a steady-state laminar flow. The latter describes the dispersion of the averaged RTD.

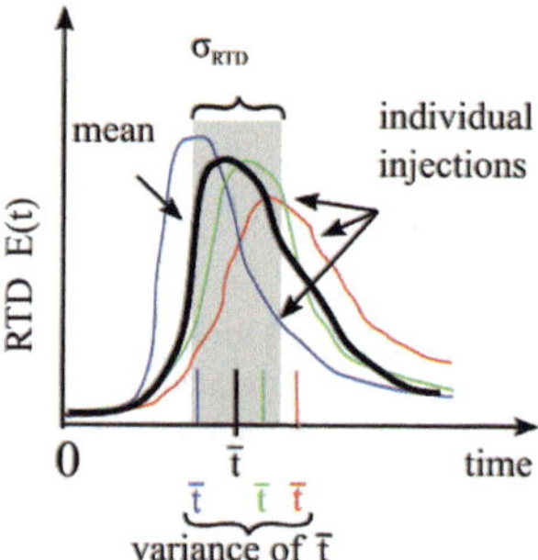

Fig. 5.9: Difference between the residence time variance σ_{RTD} and the variance of the mean residence time.

The skewness s_{RTD}, which is a measure for the deviation from a symmetrical distribution, is very similar for all three operating conditions and indicate that the RTDs are not symmetrical but right-skewed. The Bodenstein-numbers of all three operating conditions are neither close to zero nor particularly large, which means the flow can be described by neither an ideal well-stirred reactor (WSR) nor an ideal plug-flow reactor (PFR) alone. The Bodenstein-number is below the threshold value of 7 which, according to Hagen [138], represents a transition between the two regimes. This indicates that in the flows of all three operating conditions the effects of mixing by a WSR are more pronounced than a delay by a PFR.

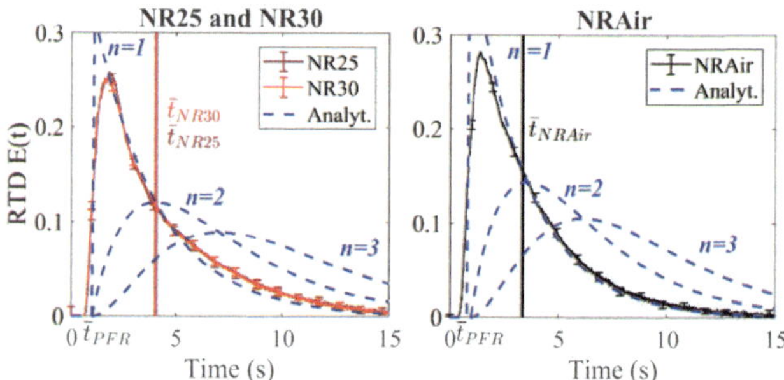

Fig. 5.10: RTD of the non-reacting operating conditions and comparison with analytical models.

Fig. 5.10 shows the measured RTD of the three operating conditions. The NR25 and NR30 operating conditions are almost indistinguishably similar, whereas the NRAir flow shows an earlier arrival of tracer and a reduced variance. The function is indeed right-skewed. However, all three operating conditions have similar RTDs, indicating that the flow patterns are similar.

For a cascade of *n* ideal WSRs of identical mean residence time in series with a PFR, the RTD can be analytically calculated by [138]

$$E(t) = \frac{1}{(n-1)!\,\bar{t}} \left(\frac{t - \bar{t}_{PFR}}{\bar{t} - \bar{t}_{PFR}} \right)^{n-1} \cdot \exp\left(-\frac{t - \bar{t}_{PFR}}{\bar{t} - \bar{t}_{PFR}} \right) \qquad \text{(Eqn. 51)}$$

A plug-flow reactor delays the flow by $\bar{t}_{PFR}$ in terms of a Heaviside-Function $\Theta(t-\bar{t}_{PFR})$. In this work, $\bar{t}_{PFR}$ is determined at the time where the tracer concentration in the initial RTD slope reaches half of the maximal value. Then, the WSRs have a mean residence time $\bar{t}_{WSR}$ of $\bar{t}_{WSR} = \bar{t} - \bar{t}_{PFR}$, with $\bar{t}$ the mean residence time of the complete reactor series.

In fig. 5.10, the analytically calculated RTDs of these combinations are shown for up to three WSRs, with the mean residence time $\bar{t}$ taken from table 5.2. The measured RTDs agree very well with a simple combination of one PFR and one WSR. Table 5.2 also lists the residence times for the delay $\bar{t}_{PFR}$ and decay $\bar{t}_{WSR}$ of all operating conditions. As predicted by the Bodenstein-number, the fluid elements remain within a WSR for a longer time than within a PFR.

Differences between the measurements and the model are evident at the RTD peak, probably as the excitation peak is not singular (infinitely short) and additional WSRs of smaller volume, e.g. the quarl, remain in the flow. While the RTD of NRAir agrees well with the model even beyond the mean residence time, the NR25 and NR30 operating conditions decay slower than the prediction by the simple model. Hence, a small fraction of the flow is presumably bypassed into a second, parallel WSR with a longer mean residence time.

Even though the internal flow structures within the combustion chamber under non-reacting conditions are complex [23] [E7], the global residence time distributions agree well with a simple combination of a WSR and a PFR.

5.2.3 Reacting Conditions

Under reacting conditions, the temperature of the flue gas is significantly increased, which leads to a reduced density and lower absorption line strength of the tracer. At the same time, the mean residence time is expected to be smaller and to feature a reduced dispersion the different flows have less time to mix, which increases the tracer peak concentration during a pulse injection. In the experiment, the tracer peak concentration is raised by a factor of 2–3. Even though the reacting conditions increased radiation and possibly beam-steering, the measurement noises for both injection types and respective optical setups remain similar (fig. 5.11a).

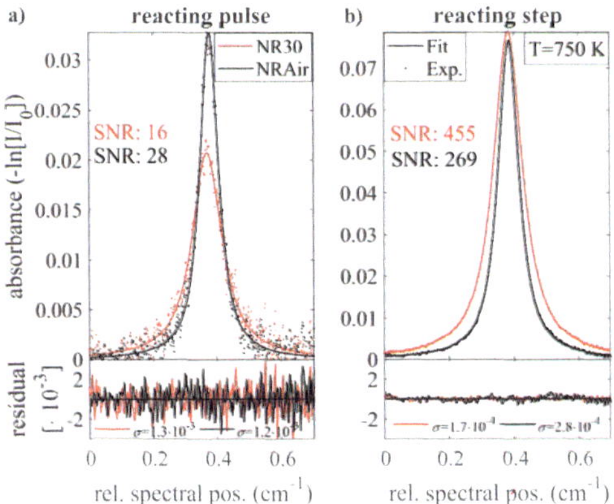

Fig. 5.11: Absorption spectra of tracer HCl under reacting conditions at peak concentrations: a) Pulse injection, HCl peak concentrations: ~158 ppm (R30)/ ~150 ppm (RAir). b) Step injection, HCl peak concentrations: ~0.23 % (R30)/ ~0.16 % (RAir). [31]

---

[31] The concentrations were calculated for a constant flue gas temperature of 750 K. This value was measured non-simultaneously using two-line thermometry, as described in the next section.

The tracer concentrations at the outlet caused by a step injection closely resemble those obtained during non-reacting conditions, as the inlet concentration is the same and the number of non-tracer molecules is not significantly changed throughout the combustion of methane. The reduction of density and absorption line strength, due to an increased temperature, decreases the SNRs by a factor of 3 for both pulse and step injections under the RAir condition. The SNRs of the R30 condition, however, are very similar to those under non-reacting conditions, probably because the high temperatures lead to a temperature-narrowing of the collisional broadening. This an increase of the line peak compensates for the loss of absorption line strength.[32]

Thermal expansion under reacting conditions decreases the mean residence times (table 5.2). In contrast to the non-reacting situations, the fired oxy-fuel operating conditions exhibit different RTD-features: the mean residence time is slightly longer for R30 than for R25, even though, as will be seen in the next section, the gas temperatures and thus the expansion of R30 are higher. Presumably, recirculation zones in the flow are reduced, which allows the tracer to flow through a larger cross-section and reduces the velocity. This hypothesis is supported by the RTD variance, which is lower for R30 and indicates a reduced stirring. Similar to non-reactive conditions, the mean residence times and the variances are lower for RAir operation compared to the respective oxy-fuel operating conditions. The Bodenstein-number has sharply increased, indicating a more pronounced plug flow behavior. Yet, it remains still within the transitional region close to the threshold, but with a higher relevance of stirring. The same applies to the oxy-fuel conditions, even though the effect is less pronounced. The RTDs of all operating conditions remain right-skewed.

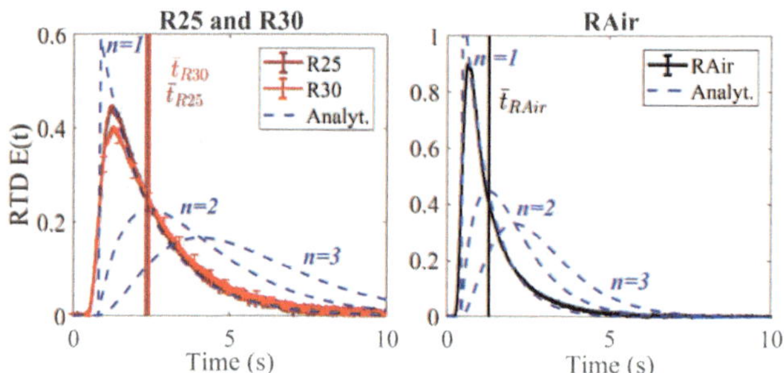

Fig. 5.12: RTD of the reacting operating conditions and comparison with analytical models.

Even though the RTDs of the two oxy-fuel conditions are now separable, their mutual similarity is still greater than compared to RAir (fig. 5.12). All RTDs can again be described by a series of one well-stirred and a plug-flow reactor with satisfying agreement. However, the relative contributions of the reactors to the overall mean residence time have changed. Under non-reacting conditions, the well-stirred reactor residence time, $\bar{t}_{WSR}$, is typically 3 times larger than the plug-flow residence time $\bar{t}_{PFR}$. Under reacting conditions, this factor is reduced to two, which, similarly to the Bodenstein-number, suggests a reduced relevance of stirring. Even though the overall agreement with the analytical WSR-PFR series is high, all operating conditions show deviations at the peak concentration and the exponential decay. While the first effect is probably due to a non-singular excitation-pulse, the deviation of the latter is too long to originate from the same effect.

In this section, two methods for measuring the RTD in combustion chambers were developed and compared, which utilized a pulse and step excitation of the chamber, respectively. For the latter, a positive and negative step injection were evaluated and found to yield similar results. The pulse-injection

---

[32] According to eqn. 35, the collisional broadening Δv_C is proportional to $(T_0/T)^{n_f}$. With n_f usually in the order of 0.5, an increased temperature leads to a narrowing of the line width, which, in certain temperature ranges, exceeds the effects of an increased Doppler broadening.

method has proven its superiority in turbulent systems as it is the only method which allows to determine the RTD without filtering or modeling, even though it has an increased experimental complexity. Using 50 injections it was possible to measure the mean RTD with 1 % uncertainty. RTDs for non-reacting and reacting operating conditions were measured and compared by means of the mean residence time, variance, skewness and Bodenstein-number. Even though the flows in the chamber are complex, good agreement was found with a series of one well-stirred and one plug-flow reactor. Under reacting conditions, the mean residence time and the variances are reduced due to thermal expansion.

5.3 Evaluation and Results: Thermochemical State

In addition to information on the flow field by means of the RTD, the thermochemical state of the flue gas of the turbulent operating conditions is highly relevant, as the temperature and species concentrations influence the viscosity, density and flame structure. Through a comparison to equilibrium calculations, the state in the flue gas allows for an estimate of the reaction progress and fuel burn-out rates. For more complex CFD simulations, measurements of the thermochemical state can serve as validation data.

As the concentrations of measured flue gas species are expected to be in the range of the respective spectrometer's detection limits, all spectrometers listed in table 3.1 (with the exception of the HCl-spectrometer) are used during these experiments. The rate of burnout and fuel-slip is studied through measurements of the concentration of CH_4, whereas a quasi-simultaneously operating OH-measurement system allows for the detection of remaining reaction zones in the flue gas. The O_2-concentration is measured for an investigation of oxygen-excess, and, in combination with measurements of CO_2 and H_2O concentrations at various positions, allows us to detect the main components of the oxy-fuel flue gas. A CO measurement allows for a detection of the main pollutant and, through comparison with equilibrium calculations, an identification of incomplete reaction progresses. Measurements of C_2H_2 enables estimations on the possibility of soot formation. The species and temperatures are measured quasi-simultaneously based on the systems described in subsection 3.2.1. First, the achieved measurement uncertainties are discussed, followed by an analysis of the flue gas temperatures and species concentrations.

5.3.1 Error Analysis

Fig. 5.13 shows representative absorption profiles measured quasi-simultaneously within the spectrometers of System I and II for both the RAir (black lines) and R30 (red lines) operating conditions. Voigt profiles of multiple absorption lines for the respective species are fitted to the measured data, and allow for a calculation of the species concentrations given in the figure. The gas temperatures are obtained from the H_2O spectrometers and, under oxy-gas combustion, from the CO_2 spectrometer.

Similar to the HCl absorption line in the previous section, the oxy-fuel combustion atmosphere is broadening the absorption lines of all species due to an enhanced collisional broadening. The pressure shift has been subtracted in the spectra for a better visualization. Due to the significantly increased absorption path length, the measurement uncertainties (table 5.3) are generally significantly smaller than during measurements at the WHP (chapter 4).

The water vapor absorption at 1854 and 1392 nm is rather strong and the residual is therefore dominated by line profile deviations rather than noise. These deviations stem from errors in the spectral data, which become apparent due to the high SNR (table 5.3). Due to the low sensitivity of the two-line thermometry towards measurement errors in this temperature region, the errors on the line-of-sight averaged temperature and species concentrations are below 2 %$_{rel}$.

The residual of the CO_2 absorption profile under R30 is also dominated by line profile deviations. The strong absorbance allows for a measurement of the CO_2 temperature based on two-line thermometry with relatively low uncertainty, which, to the author's knowledge, has not been reported in literature within the NIR yet. For combustion under RAir combustion atmosphere, the CO_2-concentration and therefore the absorbance are significantly lower. Here, the path-averaged temperature cannot be calculated with satisfying accuracy. Hence, the temperature gained from the quasi-simultaneous water-vapor absorption spectrum at 1854 nm is used to infer the CO_2 concentration.

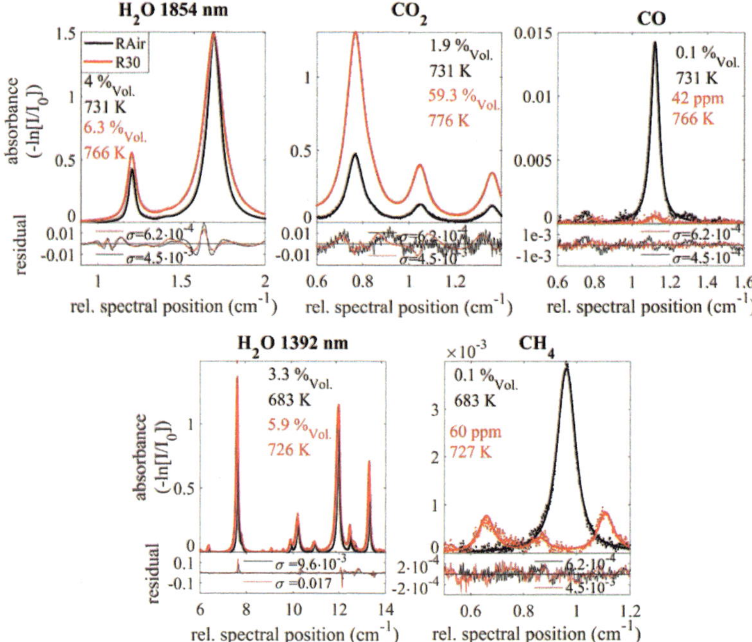

Fig. 5.13: Absorption profiles and fitted data of the species acquired with Systems I (Top) and II (bottom) for methane combustion under RAir and R30 (15 ms average). The spectrum of CO_2 under RAir has been scaled by a factor ten for a better visualization. The description within the spectra lists the concentrations of the respective species and the temperature.

It has to be noted that the absolute concentrations gained with this spectrometer showed significant deviations from known values in a heatable, constant pressure CO_2 reference cell. This behavior was linearly temperature dependent and probably stems from an inaccurate fitting of the baseline, due to an incomplete compensation of the self-induced collisional broadening. Hence, a calibration curve was derived and the concentration measurements in the combustion chamber corrected accordingly. The temperature results were in good agreement with measurements from thermocouples in the reference cell, without a need for calibration. Furthermore, at lower CO_2 concentrations of air-blown combustion and a consequently reduced impact of self-induced collisional broadening, this effect was not evident and a calibration therefore not required. The absolute concentration and temperature measurements of most of the other spectrometers were verified in chapter 4 without a need for calibration.

The residual of the CO spectrometer is influenced by line shape deviations as well as noise. The flue gas concentration of CO during R30 was below the detection limit of roughly 30 ppm and therefore not measurable. The residual of the CH_4 spectrometer during RAir combustion is purely dominated by noise with a SNR of 40. Similarly to the aforementioned concentrations of CO, the concentrations of CH_4

under R30 are below the detection limit. Instead, multiple high-temperature absorption lines of CO_2 are apparent. For all operating conditions, the C_2H_2 and OH concentrations are below the detection limit of the corresponding spectrometer.

Table 5.3: Typical detection limits and measurement errors of the spectrometers used at the Oxy-Fuel Burner with gas combustion.

System	Central wavelength target species	#		SNR	Detection limit χ_{lim}		Uncertainty	
					absolute	per length	absolute	relative
I	2314 nm CO	1	χ_{CO}	30	33 ppm	26 ppm·m	±20 ppm	±2.5 %$_{rel.}$
	2004 nm CO_2	2	χ_{CO2}	70$_{RAir}$	270 ppm$_{RAir}$	220 ppm·m	±2 %$_{oVol.}$	±2.2 %$_{rel.}$
			T_{CO2}	300$_{R30}$	0.2 %$_{oVol. R30}$		±20 K	±2.8 %$_{rel.}$
	1854 nm H_2O	3	χ_{H2O}	500	100 ppm	80 ppm·m	±0.06 %$_{oVol.}$	±1.2 %$_{rel.}$
			T_{H2O}				±3 K	±0.4 %$_{rel.}$
II	1653 nm CH_4	5	χ_{CH4}	40	60 ppm	50 ppm·m	±17 ppm	±2.5 %$_{rel.}$
	1539 nm C_2H_2	6	X_{C2H2}	<1	10 ppm	8 ppm·m		
	1527 nm OH	7	χ_{OH}	<1	10 ppm	8 ppm·m		
	1392 nm H_2O	8	χ_{H2O}	100	500 ppm	400 ppm·m	±0.08 %$_{oVol.}$	±1.9 %$_{rel}$
			T_{H2O}				±10 K	±1.3 %$_{rel.}$
III	761 nm O_2	9	χ_{O2}	85	0.16 %$_{oVol}$	0.54 %·m	± 0.4 %$_{oVol.}$	±2 %$_{rel}$

Fig. 5.14 shows the absorption profile of O_2, measured by the spectrometer of System III. The long absorption path of 3.36 m allows a high SNR even under fired operation, with a residual influenced by both white noise and profile deviations. The temperature necessary to calculate the O_2 concentration is inferred from the 1392 nm H_2O spectrometer, which is operated quasi-simultaneously to the O_2 spectrometer but along a perpendicular and shorter path (fig. 5.3). According to simulations[33] and measurements in the next subsection, the temperature at this axial position is rather homogeneous. For this reason, the temperatures along the two paths are assumed to be identical.

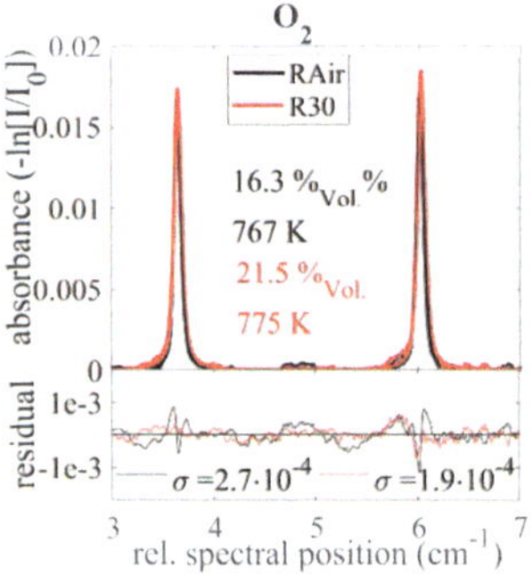

Fig. 5.14: Absorption profiles of O_2 for methane combustion under RAir and R30, acquired with the TDLAS systems III (average over 15 ms).

Table 5.3 summarizes the measurement uncertainties, which were typically about one order of magnitude smaller than the turbulence-induced variations in the process. Hence, the error-bars in the figures of the following subsections indicate the variations rather than the negligible measurement error.

---

[33] Oral communication with Samim Doost, Energy and Power Plant Technology, Darmstadt, who simulated the burner in an LES-context (unpublished work).

Time Scale Analysis

In the investigation of turbulent systems, such as the Oxy-Fuel Burner, a resolution of the smallest temporal and spatial scales (*Batchelor*-scales) is most desirable to investigate the effects of turbulence. Due to the path-integration and measurement times in the order of at least milliseconds, the TDLAS measurement system is not capable of resolving either of the scales. The integral time scale of at least the NRAir-condition, however, is in the order of 100 ms [E7]. Even though this time scale, too, cannot be measured directly due to the spatial averaging of TDLAS, effects that presumably originate from these large scale fluctuations are detectable through the measurements.

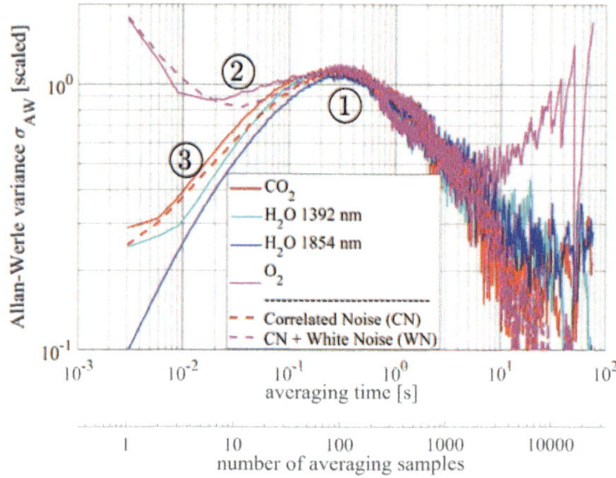

Fig. 5.15: Allan-Werle variance of spectrometers I, II and III and exponentially correlated Gaussian noise. All variances have been normalized to their value at 200 samples averaging size (R30).

In the Allan-Werle variances σ_{AW} (subsection 2.2.4) of the species concentration, calculated from the different spectrometers, an additional local maximum between the flanks of the variance (① in fig. 5.15) is evident. This maximum is identical for all spectrometers and not evident in the hook-shaped AWVs of systems without fluctuations (e.g. subsection 3.5.3). This maximum can be explained by an exponentially correlated Gaussian noise (also called red noise), which is superimposed by the TDLAS signal. For this type of noise, a concentration value is correlated to previous concentration values in an exponentially decreasing manor, reflecting the characteristics of a turbulent flow. In addition to the measured AWVs of the different spectrometers, fig. 5.15 shows a simulation of the noise with a correlation time of 0.15 s (equivalent to 50 measurement samples), calculated with the algorithm by Fox et al. [252] (red dashed line). The calculation agrees well with the Allan-Werle variances of the H$_2$O and CO$_2$ spectrometers, for which, due to the high absorbance, white noise of the measurement system can be neglected. The O$_2$ measurement, which is more severely influenced by white noise, agrees well with a simulation of a superposition of correlated and white noise of identical amplitude. An auto-correlation of the H$_2$O and CO$_2$ measurements in terms of concentration and temperature show an identical correlation time as gained from the calculation in the AWV. From these observations it can be concluded that an exponentially correlated fluctuation of 0.15 ms is evident in the process. For this reason, the averaging times of the O$_2$ spectrometer is chosen to be 30 ms (10 samples) in order to resolve the largest fluctuation but at the same time minimize white noise (②). The influence of white noise on the CO$_2$ and CO$_2$ spectrometers was found to be negligible. The signals, however, are averaged five times (15 ms) to decrease computational effort in the post-processing without significantly increasing the variance. More information on this effect and the dependence of the correlation time on the operating conditions can be found in [E6].

5.3.2 Flue-Gas Temperatures

Exhaust gas temperatures are investigated at three axial positions upstream of the combustion chamber outlet plate, 135 mm, 85 mm and 35 mm (y = 465 mm, 515 mm and 565 mm). For each axial position, multiple radial positions are investigated to determine spatial homogeneity within the central section of the burner span. At each location, 20.000 individual spectra were recorded for every spectrometer. The temperature was determined using two-line thermometry of H_2O-, and, for the oxy-fuel operating conditions, CO_2-absorption lines.

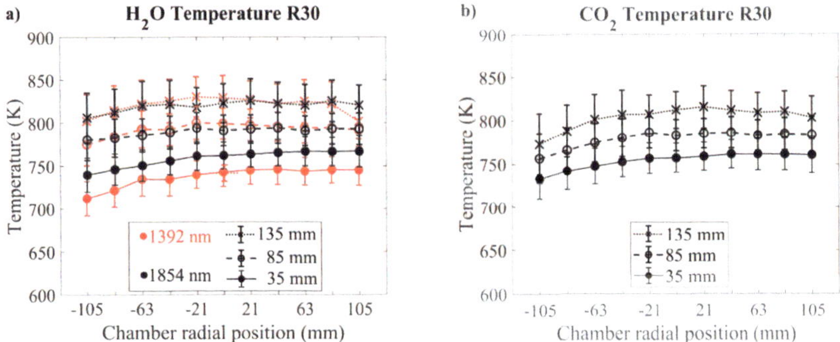

Fig. 5.16: Temperature profiles measured with the a) H_2O and b) CO_2 spectrometer (R30).

Fig. 5.16a shows radial temperate profiles based on water vapor for the R30 oxy-fuel operation condition at the three axial positions, measured sequentially using both water vapor spectrometers. The error bars indicate the standard deviations of the measurements, which were dominated by turbulent fluctuations rather than the significantly lower measurement errors. In the two upstream positions, both spectrometers show a good agreement with less than ±2 K deviation of the mean value. At the downstream position (35 mm), the deviation is in the order of 20 K. As these measurements were the first in the sequence, the difference is probably due to incomplete preheating of the chamber before measuring with the 1392 nm spectrometer.

At the same positions, fig. 5.16b shows the temperature profiles determined from the CO_2 measurements. The temperature was measured quasi-simultaneously with the 1854 nm H_2O spectrometer shown in fig. 5.16a. At the downstream position (35 mm), both spectrometers show a good agreement with a remaining difference of the mean temperature of less than 5 K, which is in the order of the spectrometer accuracy. In upstream direction, differences increase to 16 K, with the CO_2 being colder than the water vapor. This effect will be discussed below. The profiles show little dependency on the radial position, independently of the spectrometer and axial position. Hence, either the heat transfer in radial direction or the convective mixing are high. In the axial direction, a more significant temperature gradient is evident. The associated heat loss probably stems from heat transfer to the outlet plate, caused by either a recirculation zone (which were identified in the previous section) through convective transport or by radiative heat transfer. As the outlet plate is significantly closer to the measurement positions than the side-walls, the heat transfer in axial direction is higher than in radial direction.

Table 5.4 summarizes the mean temperatures and standard deviations of all three operating conditions as measured with the three spectrometers. Generally, temperatures range from 700 to 850 K, due to the globally lean conditions of the methane flame. The oxy-fuel operating conditions are slightly closer to stoichiometry and therefore have higher temperatures.

The burner and operating conditions are optimized for a numerical Large Eddy Simulation (LES). The implementation of the necessary chemical reactions and consecutive reduction through a generation of chemistry tables, however, are complex and require a large effort. In addition, the LES for this complex and large geometry with a broad range of time and length scale characteristics is time consuming and computationally expensive. Thus, the experiments are compared to equilibrium calculations with significantly lower effort in this work as a first step. These calculations are performed using Cantera [253] for constant pressure and enthalpy conditions.

Table 5.4: Line-of sight averaged temperatures, standard deviations and estimated gradients in axial direction of the three operating conditions together with flue gas temperatures by equilibrium calculations.

	Pos [mm]	$T_{H2O, 1854}$ [K]	$T_{H2O, 1392}$ [K]	T_{CO2} [K]	T_{sim} [K]	$\partial T/\partial y$ [K/cm]
RAir	135	**747.0** (±29)	**737.9** (±33)			
	85	**737.8** (±26)	**739.2** (±31)		812	-2.2
	35	**724.4** (±30)	**735.3** (±33)			
R25	135	**772.9** (±18)	**781.7** (±21)	**763.2** (±20)		
	85	**752.5** (±16)	**758.9** (±23)	**747.4** (±19)	902	-5.3
	35	**719.6** (±18)	**753.0** (±18)	**718.5** (±20)		
R30	135	**819.7** (±26)	**820.0** (±27)	**803.6** (±28)		
	85	**789.9** (±20)	**792.2** (±21)	**779.2** (±23)	904	-6.1
	35	**758.8** (±19)	**737.2** (±20)	**753.9** (±21)		

Where possible, the results were double-checked with the outlet conditions of a Chem1D [254] flamelet calculation and found to be in a good agreement.[34] Both solvers use the GRI-Mech 3.0 mechanisms to simulate the concentrations of 53 species by solving 325 chemical reactions [61]. However, it was not possible to calculate the equilibrium state for the RAir condition using Chem1D, as the equivalence ratio was outside the flammability limits and too low for the requirements of the code. Hence, the experiments are compared to the Cantera calculations in this work, which does not require flammability. The equilibrium approach is justified for the methane flame flue gas by the relatively small reaction zone compared to the combustor size [23] and the reaction rates compared to the residence time. Whether the gas flows are well mixed is debatable. In the previous section, a significant influence of mixing was identified, but it is not yet clear whether it is complete. The comparison between the measurements and the equilibrium calculations are intended to clarify whether the assumptions of this flue gas equilibrium are justified.

The chemical equilibrium calculations predict a R30 flue gas temperature of 904 K, which exceeds the experimentally measured temperature, as heat transfer to the cold walls is not considered in the adiabatic calculations. For the oxy-fuel operating conditions, radiative heat losses to the ambient are particularly enhanced through the presence of high CO_2 concentrations and an increased residence time. This becomes particularly apparent by comparing the cases RAir and R25: Even though the adiabatic equilibrium temperature of R25 is significantly higher, their measured flue gas temperatures are very similar, as the heat loss is lower for RAir.

Both oxy-fuel conditions show relatively high apparent temperature gradients in axial direction (table 5.4). This behavior presumably originates from the high CO_2 concentration and therefore increased heat transfer by radiation in axial direction. However, the measured temperatures of the R25 condition are ~ 40 K lower, even though the two operating conditions have nearly the same adiabatic flame temperature and the mean residence times were found to be very similar. Presumably, the heat loss

---

[34] The calculations were performed by Sebastian Popp (Simulation of Reactive Thermo-Fluid Systems, TU Darmstadt) using Cantera and Samim Doost (Energy and Power Plant Technology, TU Darmstadt) in Chem1D. The author kindly acknowledges the important work of the two researchers.

specifically in the near burner region is higher for R25 due to a slightly increased CO_2 concentration. In radial direction, the temperature profiles of both oxy-fuel conditions are similarly uniform and flat, due to a high radiative heat transfer in the same direction.

The RAir case, due to the significantly lower CO_2 concentrations and thus lower radiative heat transfer, shows shallower temperature gradients in the axial direction (fig. 5.17 and table 5.4). In radial direction, the RAir temperature profile is not uniform but slightly lowered ($\sim$30 K) at the centerline compared to at the radially outward positions at $r = \pm105$ mm. This effect is less pronounced at the upstream measurement positions and presumably originates from a recirculation zone upstream the outlet plate. This zone increases the convective heat transfer primarily in the center region and seems to exceed the radial heat transfer through radiation in axial direction, which is less pronounced in air atmosphere. Without line-of sight averaging along the presumably concentric temperature profile, this effect would presumably be even more distinct.

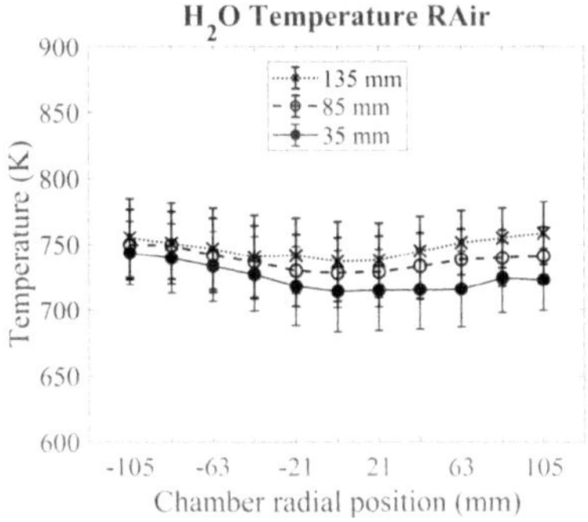

Fig. 5.17: Temperature profiles measured with the 1854 nm H₂O spectrometer (RAir).

Mixing of flame flue gas and tertiary flow

As noted, the average CO_2 temperatures of R30 (fig. 5.16) are lower than those of H_2O at the most upstream positions. This effect can not be explained by incomplete preheating of the chamber, as the measurements are performed quasi-simultaneously. The same effect was observed for the R25 operating condition, where the mean difference between the mean temperatures increases from 1 K at the outlet measurement position to 10 K at the most upstream measurement plane. This effect is counter-intuitive, as at a first glance the temperatures of the two gases should be identical at all heights.

Fig. 5.18a shows representative time series of simultaneously measured H_2O and CO_2 temperatures using for the R30 operating condition. The series were measured at the central position of both the most upstream and downstream measurement planes (top and bottom). While at the position closest to the outlet (35 mm) the two time series are in close agreement, not only with respect to their mean but also their instantaneous values, they differ occasionally (e.g. at 0.8 s and 4.3 s) in the more upstream measurement position, with the CO_2 temperature being always colder.

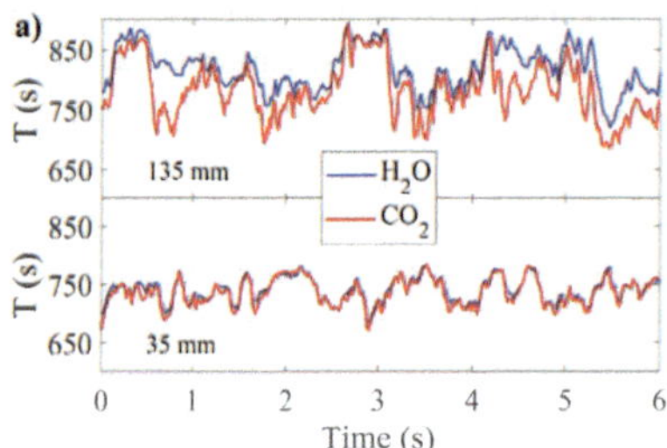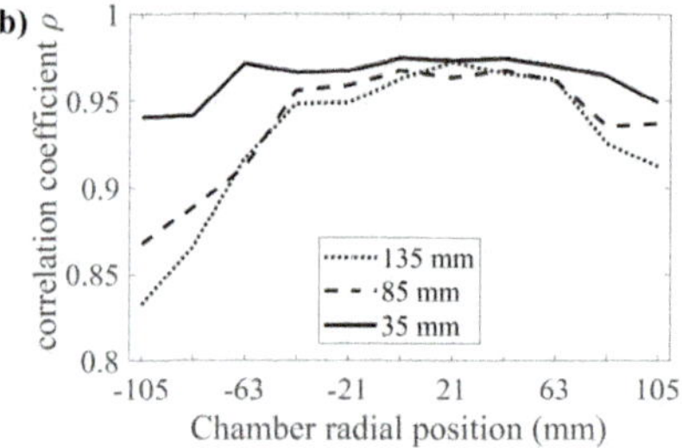

Fig. 5.18: a) Representative temperature series for the R30 oxy-fuel operating condition; b) Correlation coefficient between T_{H2O} and T_{CO2}.

Whereas water vapor originates only from combustion reactions, CO_2 is also introduced by the tertiary flow. This tertiary flow is initially cold and eventually mixes with the hot gases from the reaction zone. Inspecting fig. 5.18a, it is concluded that these gases are not well mixed in axial and radial direction at the upstream measurement plane (135 mm), and therefore the path-integrated H_2O temperature is occasionally higher than the CO_2 temperature. Further downstream, the two gases are radially and axially well mixed and therefore exhibit an almost identical temperature. This conclusion is supported by flow field measurements using particle image velocimetry (PIV) similar to those in [23], which show an intense mixing zone between the quarl flows and the tertiary flow in between the most upstream and mid measurement plane. In addition, the PIV measurements confirm that this mixing zone is spatially fluctuating, which explains the occasional agreement or disagreement between the H_2O and CO_2 temperatures.

Using the two quasi-simultaneous temperature measurements, the covariance correlation coefficient ρ is calculated. This coefficient describes a covariance between the two time series and ranges from +1 for positive to -1 for negative correlations. Fig. 5.18b shows this coefficient ρ as a function of the radial position for the three axial measurements positions of R30. The radial profile of ρ at the most downstream measurement position (35 mm) is rather uniform and close to one. Hence, the flows at this measurement plane are well mixed. The remaining deviation to one is presumably due to measurement noise. At the upstream positions, in particular at large radial positions, ρ is lower. This is attributed to incomplete mixing which results in spatially inhomogeneous gas compositions and temperatures. These inhomogeneities bias the path-integrated TDLAS measurements. However, as the overall covariance correlation coefficients are > 0.83 at all locations, the assumption of a homogeneous distribution underlying eqn. 40 still seems fairly justified.

The R25 operating condition shows a similar behavior of the correlation coefficient but with globally larger values of ρ, which indicates that the H_2O and CO_2 temperatures are more similar. This can either be due to a more complete mixing or an increased radiative heat transfer compared to R30. As the Bodenstein-numbers of both flows are almost identical, the pronounced correlation is probably due to the increased radiative heat transfer through radiation, while the mass transfer is not altered.

As the CO_2 temperature cannot be measured for RAir, the correlation and mixing are unclear for this operating condition. However, as the Bodenstein-number is higher, a further reduction of mixing is likely.

5.3.3 Flue-Gas Species Concentrations

In the following, the measured concentrations will be discussed for each species separately and compared to results of equilibrium calculations. Fig. 5.19 shows concentration profiles of the detectable species for the three measurement planes in R25. In addition, the species concentrations were averaged for each measurement plane and listed in table 5.5 for an overview of the three operating conditions.

Carbon dioxide (CO_2)

The CO_2 concentrations show a slightly decreasing profile in radial direction (fig. 5.19), probably due to the incomplete mixing with the tertiary flow described in the previous section. In the flame, CO_2 is produced and the concentration at the centerline of combustor is therefore slightly larger than at radially outwards positions, where the tertiary flow of only 75 %$_{Vol.}$ CO_2 is mixed in. In addition, a minor dependence on the axial location is evident. Nevertheless, this effect is small compared to the overall concentration and the fluctuations. The dashed line in the figure indicates simulated CO_2 concentration for a completed reaction progress. The equilibrium calculations agrees well with the experimental data, in particular close to the centerline. Due to the little produced CO_2 from the combustion compared to the amount present in the initial oxidizer (0.8 vs 75 %$_{Vol.}$), the concentration standard deviations are small compared to the mean values. Nevertheless, the standard deviations are still dominated by turbulent fluctuations rather than the significantly lower measurement uncertainties.

The measured CO_2 concentration of the other operation conditions (table 5.5) are in good agreement with the thermochemical equilibrium calculations. The flue gas concentration of CO_2 in the R25 is larger than in the R30 case due to a higher CO_2 content in the initial oxidizer feeding.

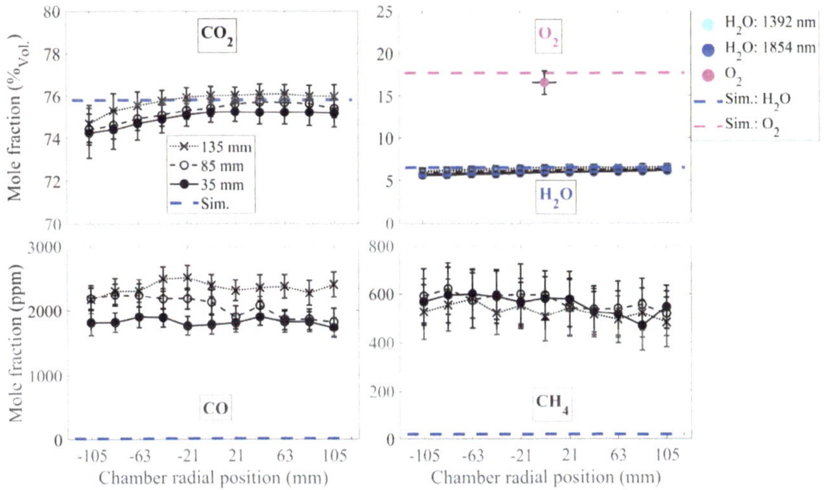

Fig. 5.19: Measured species concentration profiles for three different axial distances to the burner together with an equilibrium calculations (R25). The error bars indicate the standard deviation caused by turbulent fluctuations.

Oxygen (O_2)

Oxygen was measured path-integrated through the combustion chamber center only, as movements of the burner with respect to the sensitive optical setup let to misalignments. Fig. 5.19 shows that the O_2 concentration at the outlet (35 mm) is slightly lower than predicted by the chemical equilibrium calculations, but remains still within the 1σ uncertainty interval. Similarly to CO_2, this slightly lower mean value is probably due to remaining incomplete mixing: At the centerline, the concentrations of combustion products such as CO_2 are higher than at the radially outer positions, but lower for O_2, which is consumed during the reactions. The two measurements are therefore mutually confirming. The oxygen excess of 16.5 %$_{Vol.}$ is large, due to globally lean conditions. A reduction of the tertiary mass flow would avoid waste of precious oxygen (and CO_2) in an industrial plant, but prevent a co-firing of solid fuel within the same flame without generating globally rich conditions in the experiment.

In both the R30 and RAir case, the measured O_2 concentrations carefully match the thermochemical equilibrium calculations (table 5.5). This indicates a rather complete reaction progress of all three operating conditions, as the measured O_2 concentration would otherwise be higher due to an incomplete combustion. However, assessing reaction progress using main species concentrations is prone to errors, as deviations from equilibrium conditions below the limits of experimental uncertainties can have a large impact on minor species. Hence, an inspection of the reaction progress based on minor species is more insightful.

The flue gas O_2 concentrations of R25 and RAir are similar because the lower equivalence ratio of RAir is balanced by higher oxygen concentrations for R25. Comparing R25 with R30, a higher oxygen excess is evident for the latter, as the R30 operating condition offers both a lower equivalence ratio and a higher initial oxidizer concentration (Oxy30).

Table 5.5: Radially averaged species concentrations and standard deviations of the three operating conditions. Calc.: Results of chemical equilibrium calculations.

	axial pos.	H_2O [%$_{Vol}$]		CO_2 [%$_{Vol}$]	CO [ppm]	O_2 [%$_{Vol}$]	CH_4 [ppm]
		1854 nm	1392 nm				
RAir	135	**4.0** (±0.3)	**3.7** (±0.3)	**2.0** (±0.2)	**1070** (±90)	-	**640** (±90)
	85	**4.0** (±0.3)	**3.7** (±0.3)	**2.0** (±0.2)	**1010** (±100)	-	**580** (±80)
	35	**3.9** (±0.3)	**3.7** (±0.3)	**1.9** (±0.2)	**1300** (±320)	**16.6** (±1.5)	**570** (±80)
	Calc.	3.93		1.96	10^{-3}	16.7	<<1
R25	135	**6.2** (±0.3)	**6.4** (±0.4)	**75.8** (±0.7)	**2350** (±330)	-	**530** (±110)
	85	**6.1** (±0.3)	**6.1** (±0.3)	**75.2** (±0.7)	**1800** (±160)	-	**570** (±110)
	35	**5.9** (±0.3)	**6.0** (±0.3)	**75.2** (±0.7)	**2060** (±200)	**16.5** (±1.3)	**560** (±100)
	Calc.	6.51		75.8	3	17.7	<<1
R30	135	**6.8** (±0.5)	**6.9** (±0.3)	**71.6** (±1.3)	~40*	-	< 60*
	85	**6.5** (±0.4)	**6.6** (±0.4)	**72.7** (±1.0)	~40*	-	< 60*
	35	**6.1** (±0.3)	**6.2** (±0.4)	**72.5** (±0.7)	~40*	**22.4** (±1.5)	< 60*
	Calc.	6.42		71.0	2	22.6	<<1

*: concentration less (<) or in the same order (~) as the detection limit

Water vapor (H_2O)

Similarly to the concentrations of CO_2 and O_2, the water vapor concentrations are in good agreement to the thermochemical equilibrium calculations (fig. 5.19). The results of the two spectrometers are consistent and show a radially flat concentration profile, independently of the axial position. The other two operating conditions show similar characteristics (table 5.5). Turbulent fluctuations are again well resolvable, as the measurement uncertainty is well below the apparent standard deviation. In contrast to the O_2 and CO_2 concentrations, the fluctuations of H_2O are large compared to the mean, as this species is only forming in the reaction zone. The local concentrations are therefore highly dependent on the turbulent flow field and eventual fluctuations in the combustion process.

Consistent with the O_2 measurements, the comparison to the thermochemical equilibrium calculations indicates a rather complete reaction progress, though small deviations from chemical equilibrium are not accessible. H_2O concentrations are lower for RAir than for the oxy-fuel conditions, as the equivalence ratio is lower. The two oxy-fuel conditions show similar H_2O concentrations, as the ratio between the fuel and the gas flows of other species are similar. The higher amount of oxygen in R30 do not lead to an increased H_2O concentration compared to R25, as the operating conditions are both globally lean and have a large oxygen excess either way.

The combustion of methane produces CO_2 and H_2O in a 1:2 ratio, due to the global methane oxidation of $CH_4 + 2O_2 \rightleftharpoons CO_2 + 2H_2O$. Hence, in the RAir case where the inflowing gases contain no CO_2 or H_2O, the flue gas concentrations of the two gases exhibit a ratio of approximately 1:2. For the R30 combustion atmosphere, the H_2O concentration is slightly decreasing downstream and in return the CO_2 concentration increasing. Hence, the gases appear as not well mixed upstream. Moving downstream, tertiary flow with lower water vapor but higher CO_2 concentration compared to the combustion gases is entrained, leading to the observed reduction in H_2O and increase of CO_2. Here, the CO_2 and H_2O measurements are mutually confirming.

Acetylene (C_2H_2) and hydroxyl radical (OH)

Due to the low equivalence ratio of all three operating conditions, a formation of soot is not very likely, even for the carbon-rich oxy-fuel combustion atmospheres. Concentrations of soot precursors such as C_2H_2 are thus expected to be in the order of ppm or lower. In fact, despite the high sensitivities achieved during the measurements, C_2H_2 concentrations are below the temperature-dependent detection limits (DL) in table 5.6. This agrees with the chemical equilibrium calculations, which predict acetylene concentrations significantly below the detection limit. In addition, chemilumencence images of the flame [23] show no evidence of any soot formation.

Similarly, in spite of the high sensitivities (table 5.6), the flue gas concentrations of OH are below the detection limit, both in the mean and for all individually measured spectra. In the flame front, the OH concentration can reach values of several thousand ppm (chapter 4), which are easily detectable. It is concluded that there are no significant reactions zones remaining in the flow within the probed axial positions.

Table 5.6: C_2H_2 and OH detection limit and fuel burnout rate.

	C_2H_2 DL [ppm]	C_2H_2 sim. [ppm]	OH DL [ppm]	OH sim. [ppm]	Burnout
RAir	12	$\ll 1$	7	$\ll 1$	97 %
R25	13	$\ll 1$	7	10^{-2}	98 %
R30	16	$\ll 1$	8	10^{-2}	>99.9

Carbon monoxide (CO) and methane (CH_4)

Assessing the chemical equilibrium solely by using concentrations of major species remains limited, as the significant fluctuations prevent an identification of minor concentration differences. Measuring minor species concentrations provides a more sensitive measure.

For the R25 case, apparent CO concentrations of 1500–2000 ppm in the exhaust gas (fig. 5.19) significantly exceed the equilibrium-simulated values. According to these calculations, the CO concentration is higher for the oxy-fuel combustion atmospheres, as the CO formation is enhanced through the water gas shift reaction and others (chapters 2 and 4). The discrepancy between the experiment and the equilibrium calculations is slightly less pronounced for RAir. The CO concentrations for R30 are below the detection limit (table 5.5), and agrees with the calculations in this manner. Based on the present data, the reasons for this behavior can only be speculated.

PIV flow field studies show a strong central recirculation zone that entrains cold tertiary flow to the reaction zone downstream the quarl. Through the quick cooling by a mixing with the cold air flow, the chemical reactions are presumably quenched. Flame quenching probability is expected to decrease with higher O_2 number density, which is highest for R30. Accordingly, R30 shows the lowest CO concentrations.

High CO concentration levels are observed to be associated by high levels of unburned CH_4, indicating a fuel-slip for R25 and RAir. The associated burnout rate is presented in table 5.6. The positive correlation between CO and CH_4 concentrations may serve as another indication for flame quenching due to rapid mixing with the cold tertiary flow. The fluctuating levels of CO and CH_4 clearly exceed the measurement uncertainties. The high fluctuations observed (fig. 5.19) arise accordingly from turbulence and are similarly in line with the hypothesis that chemical reactions are terminated by mixing with recirculated tertiary flows.

The complementary CO and CH_4 measurements make the hypothesis of flame quenching by recirculation of tertiary flows plausible. However, the species were measured non-simultaneously. For a more detailed verification by correlations, the values need to be measured quasi-simultaneously, which is out of scope of this work. However, the hypothesis will be reviewed concerning the influence of the quenching to the gas-assisted solid fuel flame in the next chapter.

5.4 Conclusions

In this chapter, the thermochemical state and results of the global flow field of a turbulent methane flame under air and two oxy-fuel combustion atmospheres were investigated in detail. For studies of the global flow, the technique for measuring the RTD was significantly improved compared to literature methods. Transitions to shorter residence times and a less pronounced mixing were identified when switching from non-reacting to reacting conditions. The measured RTDs agreed well with models of a simple plug-flow/ well-stirred reactor series.

A more pronounced heat-loss was identified for the oxy-fuel compared to the air operating conditions by comparing experimental measurements to adiabatic calculations of a thermochemical equilibrium. Temperature profiles and correlations proved that the temperature distribution is homogeneous closely to the outlet, but evidences of incomplete mixing zones were found upstream. Oxy-fuel combustion atmospheres were shown to increase the temperature gradients in axial direction through a higher radiative heat transfer to the combustion chamber walls and an increased residence time. The measured concentrations of major species agreed well with thermochemical equilibrium calculations. Inspecting minor species, however, allowed a more sensitive assessment of the reaction progress. Concentrations levels of CO and CH_4 revealed that equilibrium was not achieved, probably due to partial quenching by a recirculation zone entraining cold tertiary flow into the flame. This effect decreased with increasing O_2 concentration, until it ceased for an Oxy30 combustion atmosphere.

After having clarified main features of the non-reacting and reacting conditions of methane flames, the next chapter deals with the co-firing of solid fuels in the same flames.

6 Turbulent gas-assisted Oxy-Coal Combustion: The Oxy-Fuel Burner

While the non-reacting gas-phase flow and the methane flames of the Oxy-Fuel Burner have been subject to various studies [23,255] [E1,E7,E9], the burner is primarily designed to investigate oxy-coal combustion. The combustion of solid fuels, or more precisely the co-firing of solid fuels this system is intended for, have not yet been investigated. Hence, important quantities, such as information on the flow, the gas-phase temperatures, species concentrations and burnout rate of the gaseous and solid fuels are unknown. These quantities are essential for a validation of numerical simulations [68] and to gain deeper insights into the features of oxy-coal combustion both in general and in regards to this particular burner.

The most common types of solid fuels are lignite and bituminous coal, with a global demand of more than 5 000 Mtce[35] in 2015 [2]. To fulfill the goals of the Paris agreement, the utilization of negative-emission technologies that fire biomass under oxy-fuel conditions (BECCS) seems likely to be necessary. As the chemical compositions and physical properties of biomass largely differ from conventional coals, differences in the combustion behavior are to be expected and must indeed be investigated.

The need for more detailed investigations of the oxy-coal combustion of lignite, bituminous coal and biomass drive the contents of this chapter:

- to study the practicality of the RTD measurement technique under gas-assisted oxy-coal combustion, as this method has not been applied to the hostile environments of particle loaded high-temperature environments with particles densities of up to 20 g/m^3,

- to investigate the flow field properties of gas-assisted coal combustion by means of the RTD and to identify changes when shifting from pure gas combustion to gas-assisted oxy-coal combustion. For a better separation of effects, the flames investigated in the previous chapter will be used as stabilizing gas flames,

- to measure the thermochemical state of the combustor flue gas for different combustion atmospheres (air and oxy-fuel) and solid fuels. The results are intended for a comparison with equilibrium calculations,

- to estimate the volatile and char burnouts through comparison with equilibrium calculations.

The first part of this chapter describes necessary extensions of the experimental setups and defines operating conditions for gas-assisted solid-fuel combustion. In the second section, the RTD of one exemplary solid fuel is investigated for two operating conditions. Similar to the previous chapter, the thermochemical state of the flue gas is investigated in detail within the third section and allows to draw conclusions about the reactions. Parts of this chapter have recently been published by the author in [E8] and [E10].

6.1 Experimental Setup

The combustor geometry and gas flows are identical to the previous chapter. In addition, a solid fuel feeder supplies the primary flow with pulverized solid fuel particles, which are ignited by the stabilizing

---

[35] Mtce = million tons of oil equivalent. A tonne of coal equivalent equals 7 million kilocalories (kcal) or 0.7 tonnes of oil equivalent. [2]

methane flame in the quarl. In this section, the feeder, necessary changes in the operating conditions and optical setup as well as the fuels are described.

6.1.1 Solid Fuel Feeder

A feeder located atop the burner supplies the primary flow of methane and oxidizer with solid fuel particles (fig. 6.1). For this purpose, a fraction of the primary flow is bypassed through the feeder controlled by a valve. Within the cylindrical feeder, the jet-like flow impinges onto a horizontally aligned rotating disk located at the bottom. The disk is equipped with small circumferentially arranged holes that carry solid fuel particles from the feeder compartment to the burner assembly. Particles are blown out of the holes into an outlet pipe, mixed with the remaining primary flow and dragged into the burner. The carrier holes are recharged within the feeder compartment which is filled with solid fuel particles. As the rotating disk is driven by a stepper motor with variable speed of rotation, the particle mass flow can be adjusted. The feeder is mounted onto an electronic scale balance (Soemer 1040-C3-7kg, 0.7 g resolution) to monitor the mass flow of solid fuel particles by measuring the mass loss. Since the mass flows (0.1–1 g/s) that are small compared to the net weight of the feeder, long integration times (60 s, subsection 6.3.1) are needed to measure the mass flow rate with satisfactory precision. These long integration times prevent a dynamic control of the mass flow rate, which would be necessary to compensate for short-term variations. Hence, some fluctuations of the solid fuel mass flow rate remain and increase the uncertainty in the solid fuel mass flow rate.

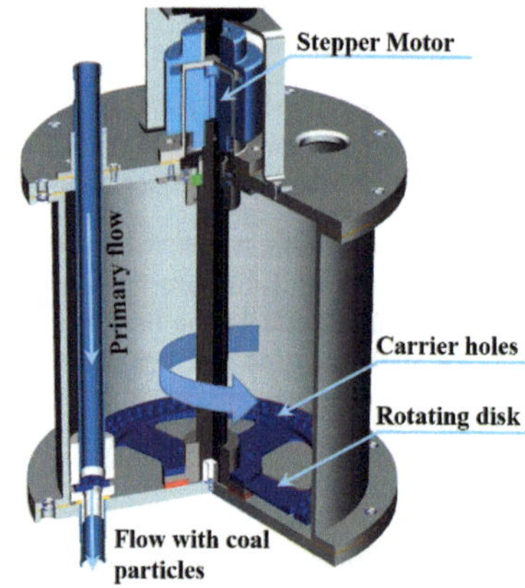

Fig. 6.1: Sectional view of the coal feeder [255]

6.1.2 Operating Conditions

During all experiments, the combustion of solid fuel particles is assisted by the identical 20 kW$_{th}$ methane gas flames of the previous chapter, which are sustained by the respective RAir, R25 and R30 gas flows and stabilized within the quarl of the burner (operating condition I in table 6.1). To study effects of particle combustion, solid fuel particles are added to the primary flow at different mass flow rates. For operating conditions OC II and OC III, the thermal powers from gaseous and particle combustion combined are 27 kW$_{th}$ and 40 kW$_{th}$, respectively.

Table 6.1 summarizes the global equivalence ratios of the three combustion atmospheres for the given operation conditions as well as the volume flows of the gas streams. The equivalence ratios are approximately independent of the fuel but strongly depend on thermal power and the combustion atmosphere. Due to the low volumetric energy density of the torrefied biomass (TBM), it was not possible to constantly support the combustion with a high mass flow rate of this fuel using the present feeder configuration. Hence, TBM was not measured at a total thermal power of 40 kW$_{th}$ (OC III).

Table 6.1: Equivalence ratio Φ of the different operating conditions.

OC	P$_{combined}$ [kW$_{th}$]	P$_{Solid Fuel}$ [kW$_{th}$]	RAir	R25	R30
I	20 [1]	-	0.19	0.27	0.22
II	27	7 RL	0.26	0.36	0.30
		7 PHBC			
		7 TBM			
III	40	20 RL	0.38	0.53	0.43
		20 PHBC		0.54	0.45

[1] investigated in the previous chapter

Compared to the methane flames of the previous chapter, the flames are elongated and their radiative intensity is increased due to particle radiation. Fig. 6.2 shows a visual comparison of the three combustion atmospheres in the combustion chamber upstream half section, with 20 kW$_{th}$ of solid fuel added.

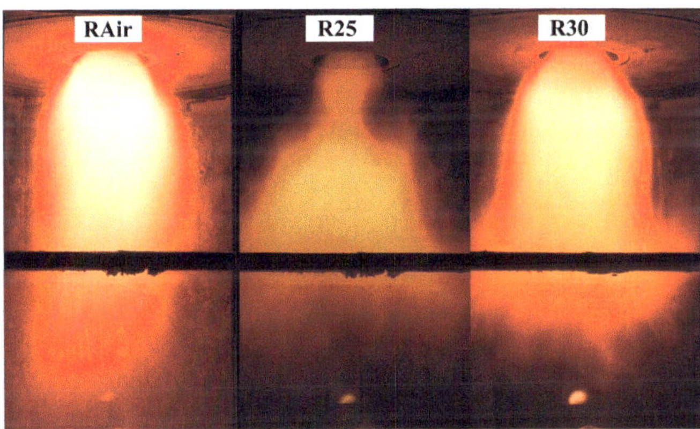

Fig. 6.2: Photography of the gas-assisted RL-flames (20kW CH$_4$/20kW$_{th}$ solid fuel, 0.2s exposure time).

6.1.3 Characterization of the Solid Fuels

Three types of solid fuels are investigated: lignite (Rhenish lignite, RL) as a representative of high volatile/low rank coal, bituminous coal (Prospher-Haniel, PHBC) as a representative of low volatile/high rank coal, and torrefied biomass (torrefied wood blend, TBM) as a representative of regenerative fuels with a similar heating value compared to lignite but higher volatile content. These solid fuels have been subject to various oxy-fuel related investigations (e.g. [31,32,53,54]). Torrefied biomass is generated by removing moisture and parts of the volatiles of raw biomass (woods, grasses) under elevated temperatures (250–400°C) within an oxygen-free atmosphere [70,256–258]. This process increases the density, heating value, grindability, and shelf-life of the biomass and makes the fuel a well-suited alternative to lignite, as it is similar in its properties but renewable and, in combination with oxy-fuel combustion, allows a removal of CO$_2$ from the atmosphere (BECCS). The torrefied biomass used in this work is produced in an industrial scale rotary drum reactor by heating a wood mixture to 295 °C for a residence time of 45 min. Afterwards, the biomass is grinded and milled.

Table 6.2: Chemical and physical properties of the solid fuels under investigation (ultimate, proximate and particle size analysis)[36].

Component			Rhenisch lignite	Bituminous coal	Torrefied biomass
Carbon	[wt%]	daf[1]	68.20	89.63	61.90
Hydrogen	[wt%]	daf	4.69	5.26	5.45
Oxygen [2]	[wt%]	daf	25.7	3.01	32.07
Nitrogen	[wt%]	daf	0.97	1.73	0.69
Sulfur	[wt%]	daf	0.44	0.37	0.02
Water	[wt%]	raw	9.13	1.08	3.72
Ash	[wt%]	dry	6.07	8.50	2.04
Volatile content	[%]	raw	44.99	28.28	62.51
Lower heating value	[MJ/kg]	daf	25.361	34.631	22.931
Higher heating value	[MJ/kg]	daf	26.376	35.750	24.113
d_{10}	[µm]		4.4	2.52	4.5
d_{50}	[µm]		29.63	14.75	15.2
d_{90}	[µm]		131.08	160.18	34.1

[1] dry, ash free; [2] from difference

Table 6.2 summarizes the chemical and physical properties of the solid fuels. While the d_{50} values of TBM and PHBC are similar, the latter has a significantly wider range in particle sizes. RL on the other hand has a similar spreading as PHBC, but a higher d_{50} value. While the hydrogen and nitrogen contents of all three fuels are similar, the RL contains considerably more water. PHBC contains significantly more carbon and has a higher heating value but lesser oxygen, water and volatiles content. All three solid fuels have a similar oxygen demand per released MJ (~1.2 mol O_2/MJ).

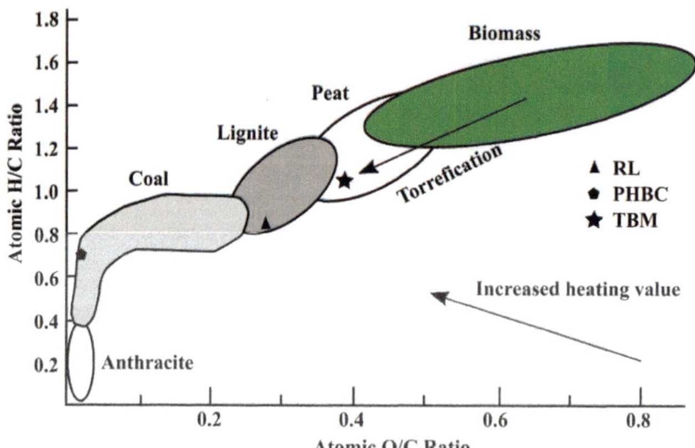

Fig. 6.3: van Krevelen diagram in style of [259] for the solid fuels used in this work in comparison with other solid fuels.

In the van Krevelen diagram [260] (Fig. 6.3), RL and PHBC are representatives for their types of solid fuel. Woods typically have an atomic O/C ratio of ~0.5–0.7 and an atomic H/C ratio of ~1.4. However, during the process of torrefaction, the biomass changes its location within the van Krevelen diagram. Due to partial devolatilization, the ratios are reduced and the chemical and physical properties of the torrefied biomass are more similar to lignite or peat. However, a higher volatile content remains.

---

[36] The solid fuels and their analyses were provided by the Institute of Heat and Mass Transfer (WSA), RWTH Aachen, Germany.

6.1.4 Chemical Equilibrium Calculations

Similar to the previous chapter, the flue gas thermochemical state is compared to chemical equilibrium calculations using the GRI-MECH 3.0 mechanism in Cantera for constant pressure and enthalpy conditions. Compared to more sophisticated simulations resolving the flow and flame, these calculations are significantly less complex and can be performed for a large parameter space of different inlet conditions. Close to the outlet, the equilibrium assumption appears justified for gas reactions (including volatiles) due to the high degree of turbulent mixing and the relatively long residence times compared to gas-phase reaction rates. However, char burnout is at timescales similar to the combustor's residence time (fig. 2.9). Thus, the conversion rate of char remains unknown. To account for this unknown, the carbon content in the equilibrium calculations is varied from 0–100% of the char content. The chemical composition of the flue gas resulting from the different calculations is compared to the experimental results and allows an estimation of the char combustion rate (using CO_2 measurements, see subsection 6.3.4).

Table 6.3: Volatile composition of the fuels after 133 ms pyrolysis.

Component [wt%]	RL	PHBC	TBM
H_2O	16.93	28.58	14.64
H_2	0.87	2.03	0.00
CO_2	26.01	4.30	9.93
CO	23.57	19.16	46.28
CH_4	4.22	13.81	6.00
C_2H_6	0.79	1.32	0.33
C_2H_4	3.77	8.17	6.90
C_2H_2	0.49	1.21	2.27
C_3H_8	23.35	21.42	13.65

The volatile composition is obtained from pyrolysis experiments performed in a drop tube reactor under pulverized fuel conditions (10^4–10^5 K/s, 1573 K) [261]. The volatile composition was derived from on-line concentration measurements by FT-IR spectroscopy (Thermo Fisher Antaris IGS USB) and gas chromatography using a thermal conductivity detector (Shimadzu GC-14B) for hydrogen only. [37]

The GRI-MECH 3.0 mechanism is primarily designed for simplified methane kinetics and therefore does not include larger hydrocarbons, tars and inorganic species. For the calculations in this work, these constituents are neglected as they contribute to the volatile mass by less than 3 %. Additionally, the fractions of propene, propyne, n-butane and isobutylene are combined and represented by propane (C_3H_8).

6.1.5 Optical Setup

The optical setup used for the investigation of gas-assisted solid-fuel combustion is similar to the previous chapter (fig. 5.3). Particle adhesion at the combustion chamber glass windows prevented a utilization of the O_2-spectrometer in the single-sided multi-pass arrangement, as the output power of the VCSEL was too low to sufficiently penetrate the particle layer sixteen times. As the operating conditions are closer to stoichiometry, the flue gas temperatures are higher and the O_2 concentrations lower, which prevented to use a lower number of passes with a correspondingly lower absorption length. For pulse-injection RTD-measurements using the HCl-spectrometer, however, the single-sided multi-pass

---

[37] The results of experiments on the volatile composition were provided by Sebastian Heuer, Department of Energy Plant Technology (LEAT), Ruhr-Universität Bochum.

arrangement was still utilizable with an increased cleaning effort. Except for the O_2- spectrometer (System III), all other spectrometers of table 3.1 are used.

In contrast to the measurements in the previous section, the HCl concentration was measured not only during RTD injection experiments, but also during investigations of the flue gas thermochemical state, as solid fuels are known for emissions of this pollutant.

An example on why rapid measurements of the chemical state are necessary is shown in fig. 6.4, where a rapid increase of the CO concentration is observed subsequently to an unintended sudden increase of the solid fuel feeding rate. Here, a high particle density presumably triggered a partial quenching of the flame and an accompanying CO increase by a factor of 50, which illustrates the benefits of a dynamic measurement technique.

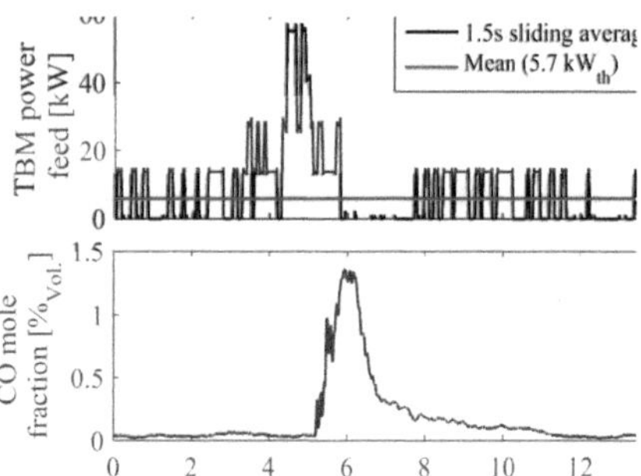

Fig. 6.4: Unintentional rapid increase of the torrefied biomass feeding rate for the TBM:R25 operating condition, followed by an increase of the CO concentration by a factor of 50.

In contrast to the previous chapter, the locations of the laser measurement paths are fixed with respect to the burner. Both cross the center of the combustion chamber in radial direction at an axial distance of 35 mm upstream the outlet plate. In solid fuel combustion, the spatial extension of the reaction zone is significantly larger and spans into the downstream half of the combustion chamber. However, as the combustion process associated to solid fuels is restricted to the close vicinity of the particles which are homogeneously distributed along the laser path approaching the combustor exit, the mixing of unburned and burned gasses in the selected measurement plane is significantly increased compared to the stabilizing gas flame. This allows similarly homogeneous conditions as for gas-phase combustion and therefore to fulfill the prerequisites of TDLAS adequately.

6.2   Evaluation and Results: Residence Time Distribution

In this section, the practicability of the developed RTD-measurement technique under solid-fuel combustion is evaluated and the results compared to the RTDs of the homogeneous combustion by the stabilizing methane flame alone. In contrast to gas combustion, coal combustion features two RTDs for the two phases, gas and solid-state particles. The particle residence time distribution is essential for the coal pyrolysis and burnout. Unfortunately, the method used in this work is only capable of measuring gas-phase RTDs. However, with a fine raw coal milling degree and high gas velocities, particle and gas residence times are said to be almost equal [73].

Compared to homogeneous gas combustion, the addition of particles into the highly turbulent quarl flow generates a hostile environment for the measurement technique. The hot particles in the beam create high-frequency background fluctuations through broadband absorption and emissions. In addition, density variations around the particles cause beam-steering due to the temperature and concentration gradients. Fig. 6.5 shows the inlet signals for non-reacting R30 in comparison with co-firing of 20 kW_{th}

Rhenisch lignite (40 kW$_{th}$ RL:R30). In the particle loaded flow, the precision of the measurement is significantly reduced due to incomplete correction of baseline-fluctuations, caused by the described effects. The mean injection signals, however, are very similar. The only difference is a quicker tracer decline after valve closing during solid-fuel combustion, which may originate in a reduced recirculation of tracer to the quarl top. Since the injection timing is not altered from non-reacting conditions, the delays between valve opening and tracer injection into the quarl were calculated from the non-reacting conditions.

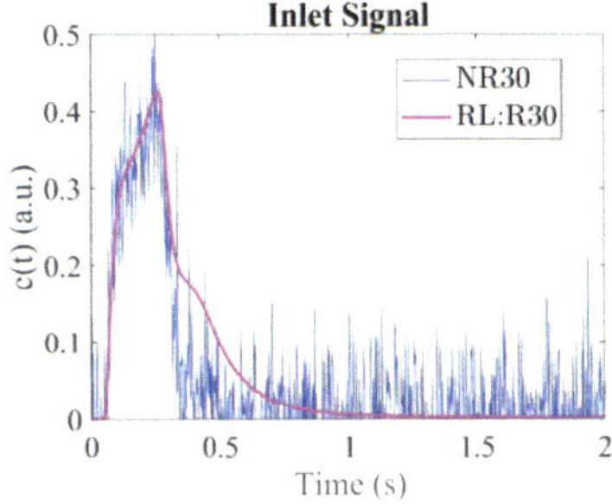

Fig. 6.5: Comparison of the measured inlet signals: NR30 and solid-fuel combustion (RL:R30 OC III, 50 injections average).

In the flue gas, the situation is less severe, as the turbulence is significantly lower, the flows temperature is homogeneous and most particles show no significant radiative emission, at least in the NIR. For these reasons, the SNR of the measurements is only mildly affected compared to the gas-flame and reduced by a factor of two (fig. 6.6).

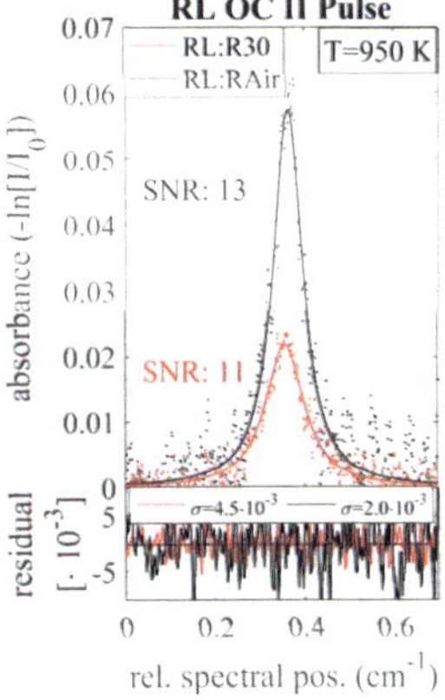

Fig. 6.6: Absorption spectra of tracer HCl under RL:RAir and RL:R30 OC III at tracer peak concentrations of ~230 ppm (R30)/ ~480 ppm (RAir).[38]

Due to higher temperatures, the absorption line strength is reduced. However, this negative effect on the SNR is balanced by increased tracer concentrations due to shorter residence times. The multi-path sensor is highly sensitive to particle-attachment on the walls. As the laser crosses the inner combustion chamber inner surfaces sixteen times, particles significantly reduce the transferred laser intensity. For this reason,

---

[38] The flue gas temperature is assumed to be 950 K, which is justified by the measurements in the next section.

the RTD of RL:RAir OC III was measured for 10 injections only, to reduce the experimental effort due to necessary cleaning of windows and cool-down. Assuming similar turbulence levels compared to the single-phase combustion, the lower number of injections causes an increased uncertainty of the mean RTD due to turbulence of 3 %.

The increased thermal expansion of the gas by the coal combustion decreases the mean residence time $\bar{t}$ further than the single-phase combustion of the methane flame alone (table 6.4). While the co-firing of RL under R30 combustion atmosphere only mildly influences $\bar{t}$, it is reduced by almost 50 % under RAir. In this case, the residence time is of similar duration as the valve opening duration, which presumably leads to an impact of the injection shape on the RTD. In fact, the RTD of this operating condition differs more severely from the combination of a WSR and a PFR than the other operating conditions (fig. 6.7), which could be either due to the described influence of the injection shape or the presence of multiple WSR within the flow.

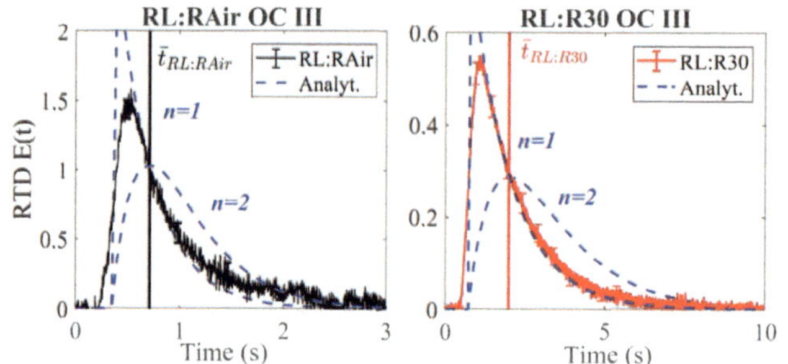

Fig. 6.7: RTD of RL co-firing in RAir and R30; comparison with analytical models

Similar to the mean residence times, the variance and skewness are reduced further (table 6.4). The Bodenstein-number of combustion under air combustion atmosphere, however, is significantly increased, either due to a formation of a more complex system of WSRs or due the systematical error by the described influence of the injection shape. As the Bo is close to the transition threshold (Bo=7), the characteristic time scales of the PFR and the WSR are almost identical.

Table 6.4: $\bar{t}$, σ_{RTD}, s_{RTD} and Bodenstein-number of RL: OC III under air and Oxy30 combustion atmosphere.

	Methane Flame (previous section)		20 kW$_{th}$ RL OC III	
	RAir	R30	RL:RAir	RL:R30
$\bar{t}$ [s]	$1.26^{\pm 0.47}$	$2.46^{\pm 0.24}$	$0.71^{\pm 0.12}$	$2.02^{\pm 0.14}$
$\bar{t}_{PFR}$ [s]	0.43	0.81	0.35	0.73
$\bar{t}_{WSR}$ [s]	0.83	1.65	0.36	1.28
σ_{RTD} [s]	$1.16^{\pm 0.48}$	$1.96^{\pm 0.17}$	$0.50^{\pm 0.08}$	$1.41^{\pm 0.09}$
s_{RTD} [s]	$2.60^{\pm 2.39}$	$1.18^{\pm 0.14}$	$0.99^{\pm 0.14}$	$1.15^{\pm 0.14}$
Bo	1.33	0.50	4.62	0.88

The co-firing of RL under R25 gas-flow and other solid fuels were not investigated, as the experimental effort was high due to the necessary cleaning of the chamber inner window surfaces. The particle absorption on the walls necessitate an improved sensor design, which either reduces the laser passes but keeps the detection limit similar by utilizing wavelength modulation spectroscopy (WMS) or reduces the particle density at the surface through purging or blow-out.

6.3 Evaluation and Results: Thermochemical State

In this section, the measurements of the flue gas thermochemical state are analyzed and compared to equilibrium calculations to obtain burnout rates. First, a post-processing strategy is defined to take the unsteady feeder characteristics into account. Secondly, the measurement errors are analyzed based on the spectra. Finally, the results of the measurements are compared to a large number of equilibrium calculations.

6.3.1 Coal Feeding Analysis

The feeding rate of solid fuel is monitored by temporally resolved measurements of the total weight of the feeder. The change in weight during the period of ~3 ms needed for TDLAS-based concentration measurement is approximately 1 mg and below the detection limit of the scale balance. Hence, the histogram of untreated feeding rates recorded at 330 Hz shows only discretized noise around zero (fig. 6.8a, blue), which requires temporal averaging in the post processing. In addition, a temporal delay between changes of the feeder characteristics and a corresponding change in the flue gas thermochemical state due to the residence time is expected and evident in fig. 6.4.

To determine suitable averaging times, cross-correlations between feeding rate and time-delayed temperatures of H_2O are calculated for varied low-pass filtering of the temporally resolved scale and temperature measurements. For the combustion of TBM (TBM:RAir OC II), fig. 6.8b shows a shallow maximum for applying a sliding average filter ranging from 30 to 60 s and zero delay between the signals. Due to the long necessary temporal averaging compared to the combustor mean residence times, a delay in between the signals does not yield an improvement of the correlation. Similar results are observed for all other operation conditions (not shown). For averaging times shorter than 30 s, cross-correlations are reduced due to noise, whereas averaging times exceeding 60 s are presumably longer than time scales of seeder-specific variations. This is supported by negative correlations observed for delays of more than ~30 s.

Applying a sliding average of 60 s and zero delay between signals, fig. 6.8a (red) shows the resulting histogram. Variations in thermal power due to fluctuations of the feeder are reasonably well resolved and are expressed by the standard variation and mean value.

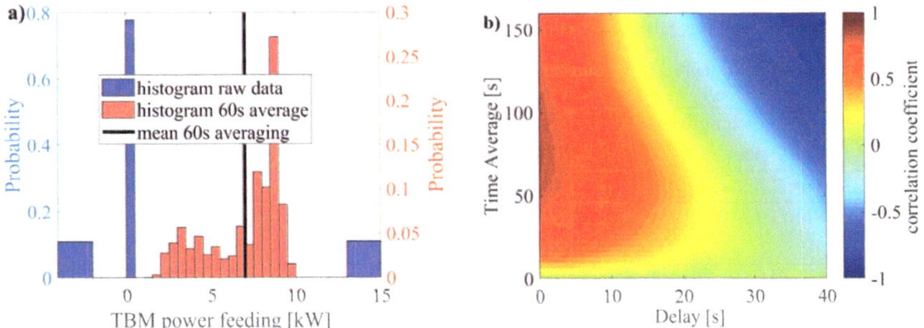

Fig. 6.8: a: Histogram of the raw and time-averaged feeding rates, b: temporal cross-correlation map between the feeding rate and the H_2O-based temperature as function of delay between the scale balance and temperature recordings and the time averaging of the signals. Both figures are exemplarily derived from the combustion of 7 kW_{th} TBM with 20 kW_{th} methane under RAir atmosphere.

6.3.2 Error Analysis

The absorption spectra allow the estimation of the measurement uncertainties and detection limits. Fig. 6.9 shows representative profiles of the combustion of 20 kW$_{th}$ RL in the RAir and R30 flame environments. The H_2O spectrometer at 1854 nm shows a signal-to-noise ratio (SNR) of 290, leading to typical uncertainties (1000 ppm/ ±5 K) and detection limits similar to those in the previous chapter (table 6.5). In the oxy-fuel combustion atmospheres, absorption lines are broadened due to enhanced collisional broadening by CO_2. Nevertheless, the uncertainties remain similar.

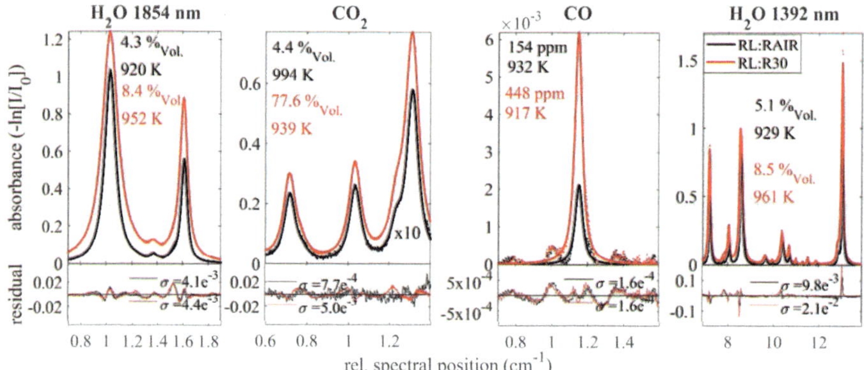

Fig. 6.9: TDLAS absorption spectra and fitted data for RL:RAir (black) and RL:R30 (red) in OC III. From left to right: profiles of H2O at 1854 nm, CO2, CO and H2O at 1392 nm. The RL:RAir spectrum of CO2 is scaled for a better visualization. Averages over 150 ms (CO) or 3 ms (others).

The uncertainties of the H_2O spectrometer at 1392 nm are around ±0.13 %$_{Vol}$/±30 K, which is larger than for a pure gas flame. However, this effect is mainly due to the increased absolute temperatures, as the relative uncertainties in temperature and species concentration remain similar to the previous measurements and are comparably low. For RL:RAir, the CO_2 spectrometer reaches SNRs of 60, leading to concentration uncertainties of 0.1 %$_{Vol}$ (2.5 %$_{rel}$). Within the oxy-fuel combustion atmospheres, the CO_2 concentrations are significantly higher, leading to SNRs of 260 and uncertainties of 2.9 %$_{Vol}$ (3.7 %$_{rel}$) and ±30 K, which is higher than the temperature error from the H_2O-spectrometer at 1854 nm, but corresponds to a similar relative uncertainty than within single-phase combustion.

Table 6.5: Typical detection limits and measurement errors of the spectrometers used at the Oxy-Fuel Burner with gas-assisted solid fuel combustion.

System	Central wavelength target species	#		SNR	Detection limit χ_{lim}		Uncertainty	
					absolute	per length	absolute	relative
I	2314 nm CO	1	χ_{CO}	40	12 ppm	10 ppm·m	±15 ppm	±1.0 %$_{rel}$
	2004 nm CO_2	2	χ_{CO2}	60$_{RAir}$	730 ppm$_{RAir}$	610 ppm·m	0.1 %$_{Vol}$	±2.5 %$_{rel}$
			T_{CO2}	260$_{R30}$	0.3 %$_{Vol. R30}$		±30 K	±3.7 %$_{rel}$
	1854 nm H_2O	3	χ_{H2O}	290	150 ppm	130 ppm·m	±1000 ppm	±1.2 %$_{rel}$
			T_{H2O}				±5 K	±0.5 %$_{rel}$
II	1742 nm HCl		χ_{HCl}	<1	7–15 ppm			
	1653 nm CH_4	5	χ_{CH4}	<1	80–250 ppm			
	1539 nm C_2H_2	6	X_{C2H2}	<1	15–50 ppm			
	1527 nm OH	7	χ_{OH}	<1	5–10 ppm			
	1392 nm H_2O	8	χ_{H2O}	150	340 ppm	290 ppm·m	±0.13 %$_{Vol}$	±1.5 %$_{rel}$
			T_{H2O}				±30 K	±3.2 %$_{rel}$

The CO spectrometer reaches SNRs in the order of 40 with low uncertainties of 1 %rel., similar to the previous chapter. Due to the varying feeding rates of solid fuels, the standard deviations of the measured temperatures and concentrations in a 60 s smoothed average typically exceed the measurement uncertainties of the spectrometers significantly. Hence, the standard deviations of the measurements in the next subsections are dominated by limited control of the inflow conditions of the combustor and turbulence rather than the optical diagnostics.

With some exceptions of the CH_4 measurements in certain experiments (PHBC and TBM under RAir atmosphere in OC II), the measured concentrations of CH_4, HCl, C_2H_2 and OH remained all below the detection limits listed in table 6.5. As the detection limit is slightly temperature dependent, a range of limits is given. The concentration of the soot precursor C_2H_2 is probably below the detection limit because the global equivalence ratios are still lean. This hypothesis is supported by the equilibrium calculations, indicating C_2H_2 concentrations well below the detection limit. The combustion of volatiles is accompanied by high concentrations of the intermediate species OH as seen in chapter 4, but localized around heat release zones of volatile combustion primarily in or close to the quarl. As OH quickly recombines and the char combustion does not release any OH radicals, only low concentrations in the order of magnitude of the equilibrium level are expected in the flue gas, which is confirmed by the actual measurements. While self-sustained coal combustion at large scales is known to release significant amounts of the pollutant species HCl [101], the concentrations in this study remained low because of the small mass flow rates of solid fuel and the incomplete char burnout which is restricted by rather short residence times in the combustor. Low CH_4 concentrations indicate a negligible slip. This is different to observations for the pure methane flame in the previous chapter, where the gas reaction progress is incomplete due to quenching by the cold tertiary flow. Only for low solid fuel loads in the RAir atmosphere, 200–400 ppm of CH_4 are detected, indicating that flame quenching may still be present to a minor extent. This observation will be discussed in more detail in subsection 6.3.4.

6.3.3 Flue-Gas Temperatures

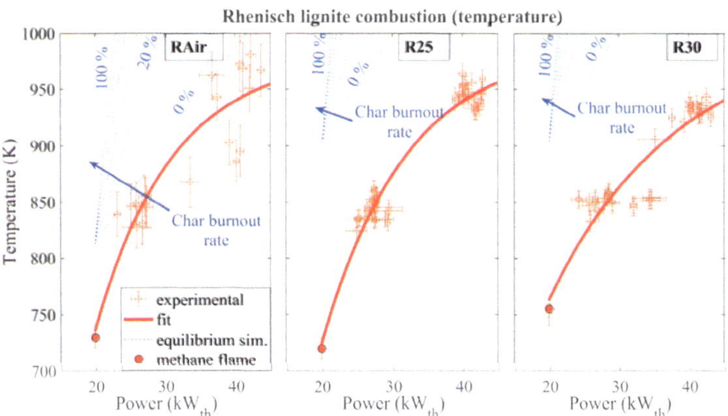

Fig. 6.10: Measured (brown) and simulated (blue) flue gas temperatures of RL-combustion in the stabilizing methane flame under RAir, R25 and R30. The red line represents a least-square fit of the experimental data to an asymptotic function.

The temperature of the flue gas is measured using the two H_2O spectrometers and, for the oxy-fuel combustion atmospheres, the CO_2 spectrometer. Fig. 6.10 displays measured temperatures of the 20 kW$_{th}$ stabilizing methane flame operated with different quantities of added Rhenish lignite. Error bars indicate the standard deviation of coal mass flow rates and temperatures during the 60 s averaging interval. For some measurements the solid fuel feeder operated more steadily, indicated by lower standard deviations and reduced scattering of respective mean values which were obtained from repeated experiments. For comparison, the temperatures resulting from equilibrium calculations for various thermal powers of the respective operating conditions are presented. The carbon content is increased in steps of 20 % from 0 % to 100 % of the fuel-specific char content to simulate varying (unknown) char burnout rates.

Independently of the combustion atmosphere, a significant temperature increase from 750 K to 950 K is observed with increasing mass flow rates of RL. For visualizing the trends, an exponential function is fitted to the data including the stabilizing pure methane flames at 20 kW_{th} (fig. 6.10, red curves). Due to convective heat transfer to the walls and radiative heat losses, which are particularly enhanced in this optically accessible combustion chamber, measured temperatures are always lower than adiabatic flue gas temperatures. For combustion in RAir atmosphere, heat losses are comparatively low and calculated adiabatic temperatures match the measurements more closely. For oxy-fuel combustion atmospheres, radiative heat losses are enhanced by higher CO_2 concentrations and longer residence times (as measured in the previous subsection), causing larger deviations from adiabatic temperatures. Deviations between measured and adiabatic temperatures increase with thermal load, because the heat transfer rate increases with higher particle densities and temperatures. At this point, from the comparison of measured and calculated temperatures no conclusions can be drawn for the char burnout rate.

The experimentally measured flue gas temperatures of PHBC and TBM combustion are presented in the appendix together with the respective equilibrium calculations (Fig. A.1). For identical nominal power levels, temperatures of the PHBC combustion are far lower than that of RL and TBM, reaching less than 850 K at combined 40 kW_{th}. This suggests that the burnout rate of PHBC is less complete compared to the other fuels, which reduces the effective thermal power. The experimentally measured temperature increase observed for each fuel is approximately independent of the operating condition. This is surprising at first glance, as the combustion atmospheres significantly differ with respect to heat transfer rate and heat capacity. However, the higher radiative heat transfer rate and heat capacity of the oxy-fuel operating conditions may have been balanced by increased equivalence ratios compared to the cases operated with air. This is different for the chemical equilibrium calculations where heat losses are not included. As such, adiabatic flue gas temperatures of the oxy-fuel operating conditions are significantly higher than for air combustion.

6.3.4 Flue-Gas Species Concentrations

Like the temperatures, the measured mole fractions are compared to equilibrium calculations with different carbon contents from the char. Because absolute values are sensitive to initial conditions and trends of concentrations depending on thermal power are more reliable, experimental and simulated results are compared primarily based on the slope of the data.

Water vapor (H_2O)

Using the two H_2O-spectrometers, flue gas water vapor concentrations are measured (fig. 6.11 for RL and Fig. A.2 in the appendix for PHBC and TBM). Due to an unsteady feeding rate of the coal at high thermal loads, uncertainties in both thermal power and measured concentrations are high, particularly for the RAir operating condition. During oxy-fuel operation, the feeding rate was more constant and the uncertainties are accordingly lower. Increasing the feeding rate of solid fuels raises the equivalence ratio and hence the water vapor concentrations. For Rhenish lignite, slopes of calculated concentrations closely match those of the measured water vapor mole fractions, considering the experimental uncertainties. The equilibrium calculations show only a minor dependence of the H_2O mole fraction on the char burnout rate, as expected, as the increase in the H_2O concentration issues from the volatile combustion only. Hence, the char burnout rate cannot be identified based on H_2O mole fractions, but the volatile burnout rate assumed to be close to or fully complete.

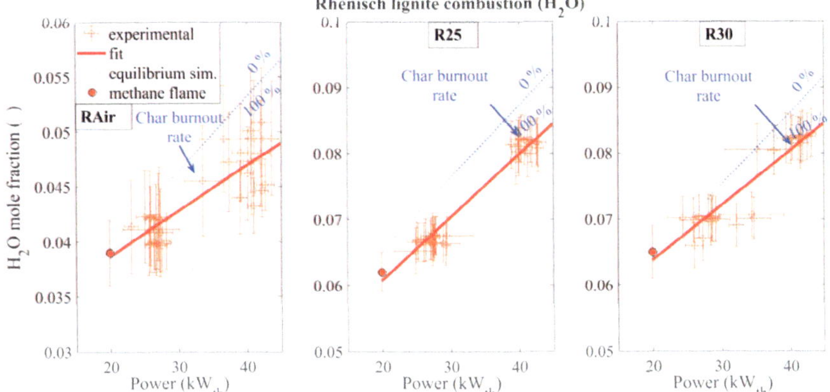

Fig. 6.11: Measured (brown) and calculated (blue) water vapor mole fractions of gas assisted RL-combustion for RAir, R25 and R30.

For combustion of TBM, independently of the operation condition (Fig. A.2), a good agreement between experimental results and the equilibrium calculations is evident for absolute H_2O concentrations and slopes, indicating a complete volatile burnout. For PHBC combustion, in contrast, the measured H_2O concentrations are lower than predicted by equilibrium calculations for all combustion atmosphere. This behavior implies an incomplete burnout of the volatiles, as H_2O is assumed to be generated predominantly during volatile combustion.

Carbon dioxide (CO$_2$)

Volatile and char combustion of solid fuels increases CO_2 concentrations in the flue gas. This can be observed in fig. 6.12 for RL and Fig. A.3 (appendix) for PHBC and TBM. For all three operating conditions of RL combustion, the fuel feeding rate was rather stable and scatter of the data is accordingly low. The equilibrium calculations reveal a strong dependence of CO_2 concentrations on the char burnout rate, which allows its estimation via comparisons between calculations and experiments. For the RAir operating conditions, the slope of the linear fit of the experimental data (red line) agrees well with a char burnout rate of approximately 50 %. Hence, for this fuel and operating condition, the char burnout appears to be incomplete. Longer residence times in high temperature zones of the combustor or an increased combustion chamber temperature due to reduced heat losses would probably promote a higher char burnout rate. For the oxy-fuel operating conditions, the burnout rates increase to approximately 80 %. This trend may be explained by a higher flame temperature (equivalence ratio closer to stoichiometry), an increased radiative heat transfer rate between the gas phase and particles due to higher CO_2 concentrations, a higher oxygen availability and a longer residence time due to lower gas volume flow rates. For high thermal powers, the CO_2 concentrations in the adiabatic calculations are presumably reduced due to formation of CO from CO_2 at high temperatures. This is in contradiction to the experiments, where heat losses cause lower temperatures than adiabatic and the decrease of CO_2 concentrations is therefore not evident.

The flue gas CO_2 concentrations of PHBC (Fig. A.3) show little scattering. The slope for RAir conditions correlates with the calculations for char burnout rates ranging from 20 % to 40 %. Hence, the burnout rate of this fuel appears significantly lower compared to RL for all three operating conditions. TBM combustion for RAir and R30 combustion atmospheres, however, shows a significantly enhanced burnout rate compared to PHBC and even exceeds that of RL for R30. This observation is probably due to the lower char content and lower particle diameters of TBM compared to PHBC. A lower char content increases the volatile and therefore near-burner combustion, which in return increases flame stability and gas temperature. Lower particle diameters increase the surface-to-volume ratio of the particles, and this effect enhances the char reaction rate through an increased gas-particle radiative heat transfer and a higher oxygen availability at the particle surface. The slope of the CO_2-increase for TBM under R25 is higher than predicted by the equilibrium calculation, even for a complete char burnout rate. Any reasonable explanation for this effect remains open.

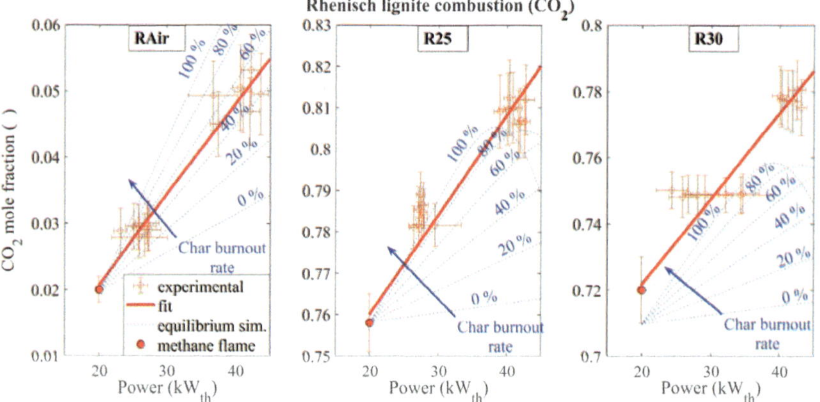

Fig. 6.12: Carbon dioxide mole fractions of the RL-combustion in the stabilizing methane flame under RAir, R25 and R30.

Table 6.6 summarizes the approximate char burnout rates of the different solid fuels and operating conditions. In general, char of PHBC has a lower burnout rate than the other fuels, and the oxy-fuel operating conditions enhance the burnout. Important factors for increased char burnout are a higher flame temperatures due to a higher equivalence ratio closer to stoichiometric conditions, an increased heat transfer between gas phase and char particles and an increased residence time in the flame.

Table 6.6: Char burnout rate estimation

Component	RL	PHBC	TBM
RAir	50 %	20 %	50 %
R25	80 %	40 %	100 %
R30	>80 %	40 %	100 %

Carbon monoxide (CO)

In the previous chapter, it was observed that CO concentrations produced by the stabilizing methane flame significantly exceeded the equilibrium calculations for RAir and R25. This effect was explained by severe quenching caused by the cold tertiary flow before reactions reached a chemical equilibrium. It was noted that this effect may decrease at higher thermal power due to a lower influence of the tertiary flow and a faster reaction progress. Inspecting fig. 6.13 for these operating conditions, the CO mole fraction is indeed decreasing at higher solid fuel feeding rates for a feeding of 7 kW_{th} of solid fuel. Nevertheless, this effect is ceased for OC III: RAir (20 kW_{th} solid fuel). For R25, the CO concentration even starts to increase again for 40 kW_{th} thermal power. This effect is in good agreement with the equilibrium calculations that predict increasing CO concentrations due to a higher equivalence ratio and higher gas temperatures. For high char burnout rates, the equilibrium calculation predicts high CO concentrations due to the Boudouard-equilibrium. The experimental data reveal CO mole fractions that exceed the values expected from a 50 % char burnout rate for RAir as concluded from evaluating measured CO_2 concentrations (table 6.6). This implies remaining flame quenching by the tertiary air flow.

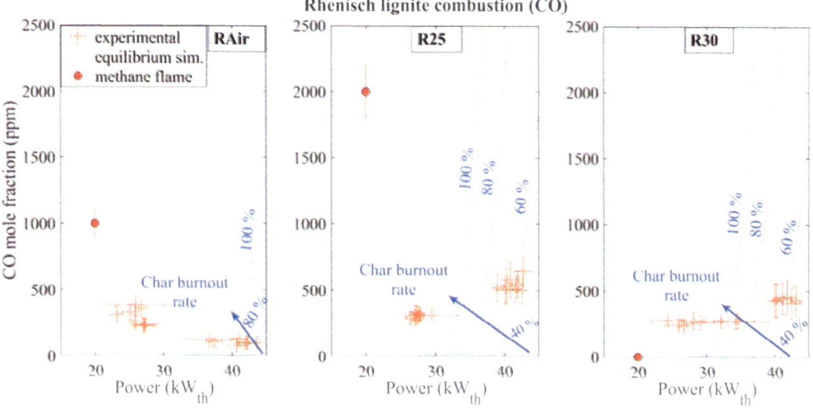

Fig. 6.13: Carbon monoxide mole fractions of gas-assisted RL combustion for RAir, R25 and R30.

For the R25 and R30 combustion atmospheres at high thermal power, the measured CO concentrations are lower than what would be expected for an 80 % burnout rate as estimated from the CO_2 measurements and equilibrium calculations (table 6.6). In particular for R30, flame quenching through the tertiary flow is less pronounced and the differences may be dominated by significant heat losses at elevated temperatures that are not accounted for in the equilibrium calculations.

The opposite of this observation holds true for the CO concentrations of PHBC combustion under the oxy-fuel operating conditions (appendix, Fig. A.4). Here, the measured concentrations exceed the equilibrium calculations for the respective char burnout rates at higher solid fuel feeding rates. This effect most likely stems from the poor char combustion progress and therefore a locally remaining CO-enriched atmosphere around the char particles. These effects may cause increased spatial inhomogeneities that could explain the severe scattering of the measurement points, as the CO concentration would be sensitively dependent on the reaction progress. For low particle densities in the RAir and R25 combustion atmosphere, however, the CO concentration is lower than for the assisting methane flame alone because flame quenching is reduced. For TBM combustion under all combustion atmospheres and PHBC/RAir, the CO concentration is decreasing with increasing thermal load in the investigated fuel feeding range, as the effects of equilibrium-CO only play a role for higher thermal loads.

6.4 Conclusions

The residence time distributions and flue gas thermochemical states of various gas-assisted solid fuel flames under air and two oxy-fuel combustion atmospheres were measured and compared to reactor flow model and equilibrium calculations.

It was demonstrated that the previously developed measurement technique is capable of measuring residence time distributions and higher statistical moments under pulverized fuel combustion. However, a further reduction of the residence time requires shorter injection pulses, as the influence of the pulse shape on the RTD may not be neglected any more. Compared to a homogeneous methane flame, the residence times and variances of the heterogeneous solid-fuel/gas flames are reduced. In terms of the RTD, the flow can globally still be described by a combination of a well stirred and a plug flow reactor with satisfying agreement.

A comparison to equilibrium calculations using Cantera to the measured thermochemical states showed reasonably good agreement with respect to water vapor concentrations, which allowed to assume a high or complete reaction progress of the volatiles. The CO_2 concentrations and their dependence on fuel and thermal load were explained by differing char burnout rates deduced from the equilibrium calculations. While Rhenish lignite and torrefied biomass, due to their high volatile content giving rise to strong volatile combustion, revealed a high char burnout rate, the burnout rate for Prosper Haniel bituminous coal was poor. Oxy-fuel combustion atmospheres, due to their higher carbon dioxide contents, increased gas-particle radiative heat transfer and longer residence times, boosted burnout rates. Considering the limitations of the equilibrium calculations and particularly the negligence of heat loss mechanisms and flame quenching, trends between measured and calculated CO concentrations agreed rather well.

In the next chapter, the combustion of a pulverized solid fuel without a stabilizing methane flame but with a complete char burnout is investigated in air and oxy-fuel atmospheres.

7 Turbulent Oxy-Coal Combustion: The WSA-*Technikum*

To stabilize a pure oxy-coal combustion and fully burn the char without the need for an assisting oxy-gas flame, larger facilities are required with reduced heat losses. The Technikum at the RWTH Aachen (Institute of Heat and Mass Transfer, WSA) offers such an environment. The combustion chamber of this facility is insulated and significantly longer than the Oxy-Fuel Burner, which enables a self-sustaining coal flame through a higher residence time in hot environments. The combustion chamber diameters of the Oxy-Fuel Burner and the Technikum are similar (42 cm vs. 40 cm). The main differences, besides the combustion chamber length, originate from the walls: The Darmstadt Oxy-Fuel Burner offers spacious optical access, as the walls are mainly fused-silica windows. These windows cause large heat losses through radiative heat transfer and heat conduction. In the Technikum, layers of refractory bricks prevent most of these losses and allow to stabilize solid-fuel flames without the need of an assisting gas-flame. The layers, in turn, limit the optical access and impede optical in-situ measurements. Hence, optical probes were specifically developed for the laser absorption spectroscopy system of this work to allow measurements of the thermochemical state within the Technikum. Using these probes, it was possible to demonstrate measurements during combustion of Rhenish lignite, Prosper-Haniel bituminous coal and an oil heating flame.[39]

The combustion of Rhenish lignite in the Technikum has been extensively studied. Toporov et al. [262] and Heil et al. [250] developed swirl burners for a complete combustion of the solid fuel under a wide range of oxy-fuel atmospheres. Hees et al. [31,32], Zabrodiec et al. [263] and Habermehl et al. [54] studied the flow fields, species concentrations, particle and wall temperatures and the reaction zone shape using LDA, FTIR-spectroscopy, two-color pyrometry and chemiluminescence-imaging during the combustion of Rhenish lignite in the Technikum under air and oxy-fuel atmospheres. In the "regular" gas flow condition of the swirl burner, the flame is characterized by a cone-shaped swirling combustion zone with a distinct inner recirculation zone, which is breaking down during variations of the near-burner equivalence ratios.

However, the facility also allows the combustion of renewable solid fuels, in particular torrefied biomass, under air and oxy-fuel atmospheres, which is of interest for BECCS. The combustion of this type of fuel has rarely been studied in-situ in either the Technikum or comparable facilities. Smart et al. [264] investigated the heat flux and flame luminosity of shea meal co-firing in coal flames under oxy-fuel combustion atmospheres at a larger facility. Skeen et al. [265] measured nitic oxide emissions in a 30 kW_{th} oxy-fuel combustion facility during combustion of coal/saw dust mixtures. Möller et al. [266] presented narrow-band emission images of torrefied biomass combustion in the Technikum under air atmosphere. The author of this work measured H_2O, HCl and CO_2 profiles for the combustion of torrefied beechwood [E11].

The need for more detailed in-situ investigations of renewable-fuel combustion under oxy-fuel atmosphere in close-to-application facilities defines the aims of this chapter:

- to study the applicability of the specifically developed probes under the combustion of biomass,

- to measure the thermochemical state of torrefied biomass combustion under air atmosphere at various axial distances to the burner for a variation of the near-burner equivalence ratio,

- to compare the thermochemical state of air-blown combustion to the combustion under oxy-fuel atmospheres with different oxygen content and the results of both to equilibrium calculations,

---

[39] unpublished work

- to identify changes between the oxy-fuel combustion of TBM and RL through the comparison with previous measurements under similar operating conditions.

The probes used in this chapter have already been presented by this work's author in [E11].

7.1 Experimental Setup

First, the Technikum and the investigated operating conditions will be described in detail. Afterwards, the specifically designed probes and adaptions to the equilibrium calculations will be introduced.

7.1.1 WSA-Technikum

The Technikum comprises of a vertically displaceable, down-firing swirl burner atop a cylindrical combustion chamber of 400 mm inner diameter and 4.2 m total length (fig. 7.1a). Four ports at one axial height allow access to the combustion chamber and measurements using various techniques. The flue gas exits the combustion chamber through a central aperture at the bottom of the combustion chamber where it is quenched and cooled down by means of a water injection.

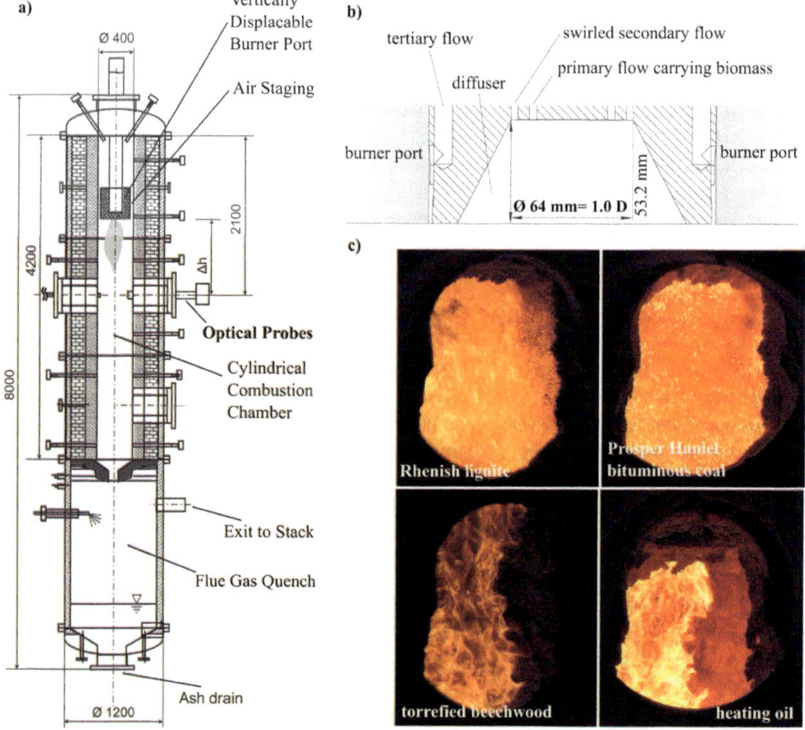

Fig. 7.1: The Technikum: a) vertical cross section through the burner (all dimensions in mm); b) cross section through the burner [31]; c) photography of exemplary pulverized solid fuel flames (and heating oil as comparison) under air combustion atmosphere [40].

---

[40] Provided by Institute of Heat and Mass Transfer, RWTH Aachen

The burner head (fig. 7.1b) is similar to the Oxy-Fuel Burner in the design by Toporov et al. [250] and enables an intense backflow to the burner quarl for a fast and stable ignition of solid fuel particles and a reliable flame stabilization. Similar to the Oxy-Fuel burner, a partially premixed primary flow with solid fuel particles issues from an inner annular crevice. A major portion of the oxidizer flow through the burner is provided through the secondary annulus, which is swirled before entering the combustion chamber. Together with the primary flow, it allows for a variation of the swirl number. A third, near-burner flow enters through an additional annular gap between the burner and the burner port. The flow field and gas composition in the vicinity of the burner are predominantly governed by these gas streams. Thus, these streams are embraced together with the mass flow rate of solid fuel in the local equivalence ratio Φ_L.

An equivalent to the Oxy-Fuel Burner's tertiary flow allows air-staging. This staging flow enters the combustion through the annular gap between the combustion chamber's walls and the axially displaceable burner port. The four flows and the solid-fuel mass flow rate define the global equivalence ratio Φ_G. More details on the burner geometry can be found in [31].

Fig. 7.1c shows photography-images of the visible flame appearances in the Technikum, with flames generated by different fuels. As the torrefied wood-blend used in this work has a composition very similar to torrefied beechwood, the flame appearance can be assumed to be likewise similar. The latter flame, due to high volatile content and low particle diameter, seems to have a closed flame surface area with a structure more similar to an oil flame rather than a coal flame.

7.1.2 Operating Conditions and Equilibrium Calculations

Three air-blown operating conditions with variations of the near-burner (local) equivalence ratio Φ_L are investigated. The "regular" condition is locally slightly fuel-rich ($\Phi_L = 1.25/ \lambda_L = 0.8$), but globally lean ($\Phi_G = 0.78/ \lambda_G = 1.3$). Deviating from this standard, the global equivalence ratio is kept identical but Φ_L moved into a more fuel-rich regime ($\Phi_L = 1.67/ \lambda_L = 0.6$) for one flame and into a lean condition ($\Phi_L = 0.91/ \lambda_L = 1.1$) for a second flame variation. The latter flame requires the majority of the combustion air to pass through the quarl. These operating condition are all termed RAir, and subdivided in the regular, rich and lean flames.

The combustion atmosphere of the regular operating condition is replaced by two different oxy-fuel atmospheres, Oxy21 and Oxy25, in the additional operating conditions R21 and R25. An equal ratio between the axial and tangential flow velocity of the secondary stream (swirl rate), has been employed during all experiments. The fuel, torrefied biomass, is of the identical batch as during the experiments in the previous chapter (table 6.2). Table 7.1 summarizes the operating conditions and their respective flow rates.

Table 7.1: Operating conditions under investigation: Equivalence ratio, oxidizer, power and volume flows.

	Air			Oxy-Fuel	
	RAir	RAir	RAir	R21	R25
Local equivalence ratio Φ_L (primary to tertiary stream)	1.67	1.25	0.91	1.25	1.25
Secondary stream swirl number			1		
Global equivalence ratio Φ_G			0.87		
Power (kW$_{th}$)			60		
Flow Volume					
Primary flow (m$_3$/h)	8.5	7.6	8.5	8.5	7.6
Secondary flow (m$_3$/h)	21.1	25.3	44.7	44.7	25.3
Tertiary flow (m$_3$/h)	3.7	4.5	7.9	7.9	4.5
Staging flow (m$_3$/h)	38.9	23.3	11.1	11.1	23.3

Similar to the previous chapter, equilibrium calculations of the operating conditions are performed using Cantera. Within these calculations, char is simplified to be purely composed of carbon. In this approach, the results of the calculation were not meaningful, with temperatures and CO concentrations being far too high to be physical. The author's personal presumption, without being an expert in kinetics, is that the reaction rates and corresponding enthalpy releases of elementary carbon in the GRI-MECH 3.0 mechanism feature large errors. This is insignificant for the description of methane combustion processes the mechanism is intended for, where elementary carbon concentrations are negligible. However, due to the large contents of carbon in the Technikum, the errors result in unphysical solutions.

As a solution to this issue, the elementary carbon in the calculations was replaced by the same amount of CO, which is justified by reactions through the heterogeneous oxidation reaction (subsection 2.1.4)

$$C + 0.5\ O_2 \rightleftharpoons CO + 111\ kJ/mol \qquad \text{(Eqn. 52)}$$

The O_2 concentration is reduced accordingly. To account for the enthalpy transfer, the inlet temperature is increased by $\Delta T = \Delta h / c_p$.[41] This approximation is similar to the approach in [68] and the results of the equilibrium calculations with this strategy implemented seem far more meaningful. In the equilibrium calculations of the previous chapter this approach was not necessary, as the two approaches resulted in similar results, speculatively due to the significantly leaner operating conditions.

7.1.3 Optical Setup

The measurement system heavily relies on two water-cooled optical probes, which were inserted into opposing optical ports of the facility (fig. 7.2). These two ports allow to establish a single-sided double-pass. The major part of the optical elements are mounted onto the rear side of one of the probes (laser probe) within a purged box. The optical system is based on the previously used insertion of a collimator into an OAP. The laser enters the internally hollow probe through a wedged window (fig. 7.3a). The pipe-type probe is water-cooled and purged by high-purity nitrogen. At the probe tip, the laser beam enters the combustion chamber through a second wedged window.

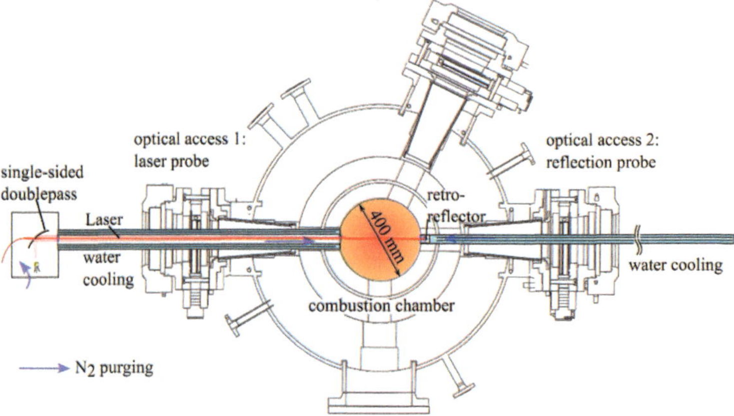

Fig. 7.2: Horizontal cross section through the Technikum at the height of the observation ports with inserted probes.

To prevent adhesion of particles, a purging flow is utilized for cleaning and separation to the combustion gases by means of a CFD-optimized film flow atop the window surface. At the opposing combustion chamber side, the laser is reflected by a fused-silica retroreflector, which is mounted to the tip of the

---

[41] The temperature-dependent heat capacity is calculated from the Shomate equations [267] of H_2O, CO_2, O_2, N_2 and CO, weighted by their relative concentrations in the unburned gases.

second probe (reflector probe). During the experiments, this retroreflector became apparent to be highly resilient against the high temperatures and temperature gradients in the combustion chamber as well as the abrasion by solid particles. Similar to the laser probe, it is water-cooled and the retroreflector surface is protected by a nitrogen film flow. The probe diameter, however, is significantly smaller, as the probe is not required to be hollow. A traverse allows to move the probe into the combustion chamber and to adjust the absorption length.

a)

b)

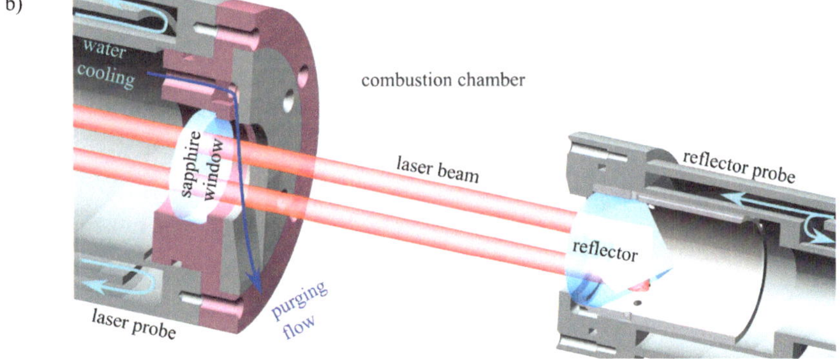

Fig. 7.3: a) Probe tips after ~24 h (laser probe) and ~3 h (reflection probe) operation under combustion of RL and PHBC. b) sectional view of the probes.

Fig. 7.3a shows the probe tips after several hours of operation. In the flue gas, the probes were operated for long times without the need for cleaning or a significant loss of laser intensity. Close to the burner, however, particles penetrated the film flow, in particular of the reflection probe, and sticked to the transparent surfaces due to the tar in the particles. This decreased the operation time without cleaning to as low as ~20 min. After cleaning, which involved scratching tar residuals off the window and reflector surfaces, the probes were fully operational again.

One of the remaining ports is equipped with a window for visual inspection, which allowed for a simultaneous narrow-band filtered imaging of the flames published in [266].[42] In these images, an inhomogeneous reaction zone along the laser path is apparent. Hence, the thermochemical state along

---

[42] The authors of the referenced work attributed the narrow band radiation (425–435 nm bandpass filter) to CH*-chemiluminescence. However, a more recent publication [268] noted the radiation to originate from two sources, CH*-chemiluminescence and particle blackbody radiation. Hence, the images mainly indicate the char burnout zone rather than volatile combustion.

this path can be assumed to be inhomogeneous as well, in particular in the near-burner region. The nonlinear influences of this effect on the measurement technique result in unknown systematical errors. In a first approximation, the measurements can be assumed to be similar to the mean thermochemical states along the path.

7.2 Evaluation and Results: Thermochemical State

7.2.1 Error Analysis

Due to limited operation time, it was only possible to perform measurements with one diagnostic system. Based on previous experiences at the Technikum with RL combustion, System I was chosen, at it promises a higher number of detectable species within the facility conditions. Fig. 7.4 shows measured spectra of the respective species in the burner far region (axial distance 6 D = 385 mm) for the regular RAir and R25 atmospheres. Due to the necessity of calibrating the CO_2-spectrometer described in section 5.3.1, it was not possible to determine the CO_2 concentrations during the high concentrations of the R25 condition. The reference cell utilized for the calibration could only be operated at temperatures far lower than the temperatures in the Technikum and an extrapolation was too sensitive to yield meaningful results.

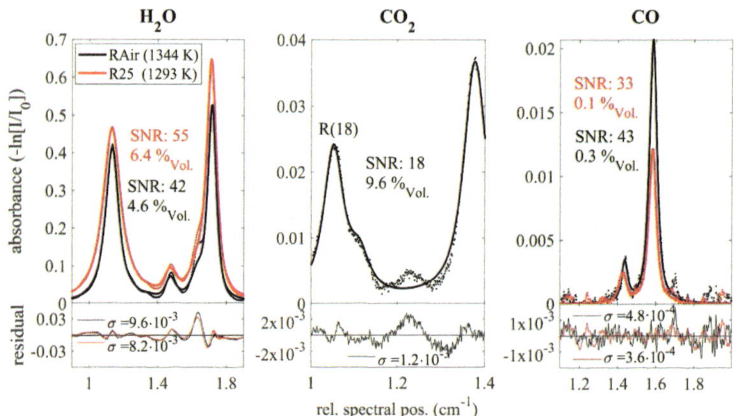

Fig. 7.4: Spectra of H$_2$O, CO$_2$ and CO in RAir ($\mathbf{\Phi}_L = \mathbf{1.25}$) and R25.

The H_2O spectrometer reveals one high-temperature absorption line which is not sufficiently accounted for in the HITRAN/HIGHTEMP databases. Hence, the SNR of the H_2O spectrometer at 1854 nm is significantly lower than in previous measurements and the uncertainties increased accordingly (table 7.2).

Table 7.2: Typical detection limits and measurement errors of the spectrometers used at the Technikum for pure TBM combustion.

System	Central wavelength	#		SNR	Detection limit χ_{lim}		Uncertainty	
	target species				absolute	per length	absolute	relative
I	2314 nm CO	1	χ_{CO}	20–50	30 ppm	24 ppm·m	±34 ppm	±3.4 %$_{rel.}$
	2004 nm CO$_2$	2	χ_{CO2}	10–20	0.61 %$_{Vol.}$·m	0.49 %$_{Vol.}$·m	0.62 %$_{Vol.}$	±6.4 %$_{rel.}$
	1854 nm H$_2$O	3	χ_{H2O}	40–60	0.13 %$_{Vol.}$·m	0.10 %$_{Vol.}$·m	±1300 ppm	±2.1 %$_{rel}$
			T_{H2O}				±16 K	±1.25 %$_{rel.}$

In the fitting process, the CO_2 spectrum has been narrowed, as a two line thermometry was not possible at the low CO_2 concentrations of RAir. Similar to the H_2O spectrometer, an unaccounted spectral feature is evident, which probably originates from one or more high-temperature absorption lines. The absorption strength of the CO absorption is high and allows SNRs similar to the previous measurements. However, due to the higher uncertainty of the temperature, the uncertainty of the CO concentration is increased.

Optimizing a trade-off between reduced noise and a high measurement repetition rates, the spectra of the H_2O spectrometer were averaged over 0.3 s and those of the other two spectrometers over 3 s. Hence, the achieved measurement frequency is considerably lower than during measurements at the Oxy-Fuel burner, probably due to the high loads of interfering particles. The error bars in the figures of the following subsections represent the standard deviation of the measurements over a measurement duration of 3–5 min.

7.2.2 Variation of the Local Equivalence Ratio

Fig. 7.5 shows average narrow-band-filtered (425–435 nm) images of the three RAir flames as measured by Möller et al. [266]. For the locally rich operating conditions, the images exhibit an elongated flame which stretches far into the combustion chamber. The flames of the locally leaner operating conditions are significantly shorter and more intense. These images assist to interpret the quantitative results of the TDLAS-measurements.

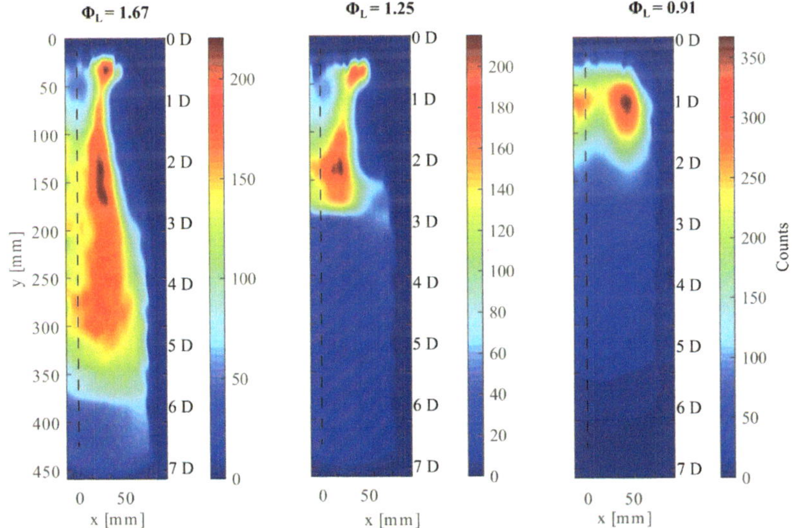

Fig. 7.5: Narrow-band emission images of the three RAir flames with different local equivalence ratio (exposure time 0.8 ms, 2 frames/s, 150 frames average).[43]

Fig. 7.6a shows the path-averaged gas temperature of the same operating conditions at various axial distances to the burner. In similarity to previous publications, the axial coordinate has been scaled with the quarl top diameter (64 mm). The temperature of the "regular" operating condition ($\Phi_L = 1.25$) is approximately constant with the burner distance. The narrow-band emission show the reaction zone of

---

[43] Images from [266], provided by Institute for Heat and Mass Transfer, RWTH Aachen. Further information on the measurement procedure and the post-processing can be found in the publication.

char combustion to be approximately located at the measurement position closest to the burner. Hence, the reaction progress at the first measurement position can be assumed to be high, and the gases in the downstream measurement positions to be mainly flue gas. The heat release by the slower oxidations, e.g. of CO, seems to balance with the heat losses in the combustion chamber further downstream. This is different to the locally lean operating condition ($\Phi_L = 0.91$). At the near burner measurement position, the temperatures of both conditions are equal. However, as the reaction zone is located closer to the burner and the remaining reactions therefore presumably cease earlier, the temperature begins to drop downstream. Furthermore, less CO is available for a downstream oxidation, due to the lean conditions near the burner. A major part of the rich flame's ($\Phi_L = 1.67$) reaction zone (fig. 7.5) is located at approximately 4–5D, which coincides with the measured maximal temperature.

The H_2O mole fractions (fig. 7.6b) show similar features, with the regular and lean operating conditions being comparable. The H_2O mole fraction of the regular condition is slightly increasing at the beginning due to a remaining volatile combustion. Due to the elongated reaction zone of the rich flame, the H_2O concentration is increasing downstream. In the most downstream measurement position, all three operating conditions have a similar H_2O concentration, as the equilibrium concentration is given by the global equivalence ratio which is identical for all three flames. The H_2O concentration of all three operating conditions are asymptotically approaching a finite value which is close to the estimation by the equilibrium calculation.

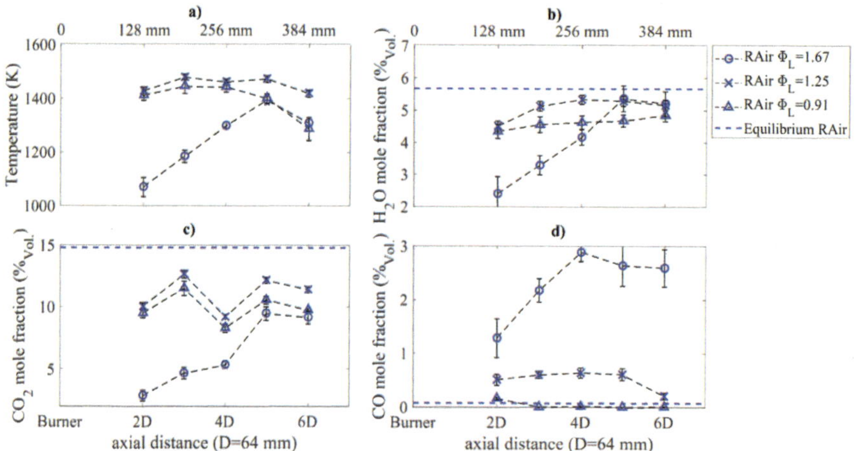

Fig. 7.6: Thermochemical state of the air operating conditions at various positions: a) temperature; b) H_2O mole fraction; c) CO_2 mole fraction; d) CO mole fraction; comparison with equilibrium calculations.

The CO_2 concentrations show larger spatial variations, but similar trends compared to the temperature and water vapor mole fraction profiles. The CO_2 release of the rich flame is elongated and the peak concentrations are reached at 5D, which is further downstream compared to the other two flames. The measured CO_2 mole fractions are notably smaller than expected from the equilibrium calculations. Whether this effect is due to incomplete char combustion or a systematical error of the spectrometer as has been evident in the measurement at the Oxy-Fuel burner remains unclear.

The locally rich conditions of the $\Phi_L = 1.67$ operating condition result in high CO concentrations at the end of the reaction zone at 5D (fig. 7.6d). Habermehl et al. [54] identified a breakdown of the inner recirculation zone under the identical operating condition for a combustion of the similar fuel RL. As the flows can be expected to be similar, this breakdown is likely to occur in the combustion of TBM as well and to be the origin of the large fluctuations, which are evident in all spectrometers for this operating condition. In addition, a large quantity of the air flow in this operation condition passes through the annular gap between the combustion chamber's walls and the axially displaceable burner port. Quenching through this cold staging flow presumably leads to high CO concentrations, similarly to the deeply staged flame configurations in chapter 5.

As the CO concentrations outside the reaction zone of the other operating conditions are significantly lower, the CO concentrations within the reaction zone can likewise be assumed to be significantly smaller compared to the locally rich flame. Downstream of the reaction zone, the CO concentration of the regular flame is slightly increasing, probably due to a formation of CO from a minor remaining char burnout. Towards the most downstream measurement position, the CO is consumed through oxidation and both flames approach equilibrium CO concentrations.

7.2.3 Variation of the Combustion Atmosphere

In fig. 7.7, the thermochemical state of the R25 and R21 operating conditions is compared to the results of the corresponding RAir flame. Due to lower heat capacities, the gas temperatures of RAir are higher than those of R25, which in turn exceed the temperatures of R21. The RAir and R25 temperatures are rather constant. However, the temperatures of R25 slightly decrease at the most downstream measurement position. This is similar to the locally fuel-lean RAir flame and allows the presumption of a reaction zone closer to the burner, due to an increased oxygen availability and lower flow velocities. The temperature of the R21 operating condition is slightly increasing towards 5D, which makes the flame similar to the locally fuel-rich operating condition of RAir. None of the measured temperatures agree with the results of the equilibrium calculations. It is not clear whether this behavior originates from heat losses in the combustion chamber or from errors in the calculations. However, the measured temperatures agree fairly well with spatially resolved temperature measurements using a suction pyrometer (Fig. A.5), which were performed under similar operating conditions (R21 and R25 with TBM replaced by RL). Hence, this behavior is not likely to be due to measurement errors.

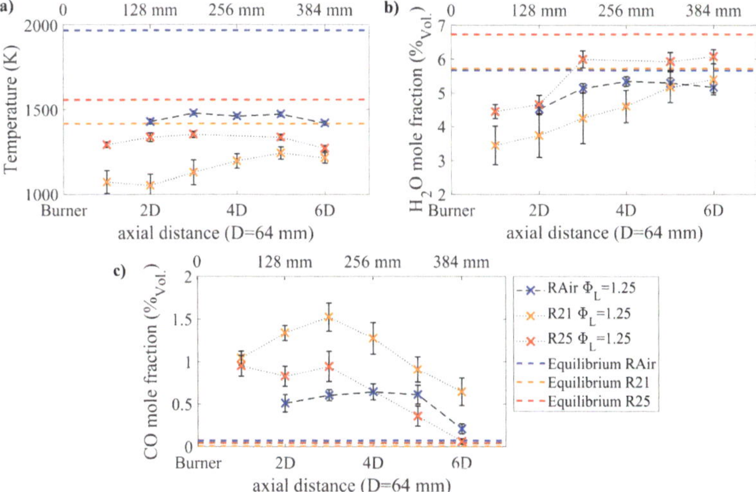

Fig. 7.7: Thermochemical state of three operating conditions with different combustion atmosphere, but identical equivalence ratio at various positions: a) temperature; b) H_2O mole fraction; c) CO mole fraction; comparison with equilibrium calculations.

The H_2O mole fractions of the R25 and RAir operating conditions rise closely to the burner and are constant afterwards (fig. 7.7b). For this reason, the reactions are assumed to be finalized at 3D and an equilibrium state to be reached. Similar to the temperature, the water vapor concentrations of R21 increases slowly, which supports the hypothesis of a combustion similar to the rich RAir flame. While the concentrations of the RAir and R21 conditions at 6D closely match the result of the equilibrium calculation, the concentration of R25 is approximately 10 % lower than the corresponding equilibrium value. A meaningful explanation for this is not evident.

In [269], the CO concentration of the R25 operating condition under the combustion of RL was recognized to exceed those of the air-blown combustion by a factor of 10. This behavior cannot be confirmed under TBM firing (fig. 7.7c), where the concentrations are similar. While the concentrations of CO for the R25 operating conditions are higher in the near-burner region, they are decreasing quicker than for RAir and approaching equilibrium conditions at 6D. The temperature and oxygen diffusivity are higher for the air-blown combustion, which should reduce the CO concentrations. However, this effect seems to be approximately balanced by a lower oxygen partial pressure. Similarly to the locally rich RAir flame, the CO concentrations are high for the R21 case. However, CO is consumed starting at 3D and at a similar rate as for R25.

Generally, the CO concentrations in the combustion of TBM are approximately twice as high as for RL under the identical operating conditions, which can be attributed to the doubled CO content in the volatile composition (table 6.3) and a generally higher volatile content of TBM. The temperature amplitudes are similar, but the water vapor concentrations slightly higher for RL [31], probably due to a higher hydrogen and moisture content in the raw fuel. These features promise to allow a similar combustion behavior of TBM compared to RL in larger facilities, however, higher CO concentrations have to be expected before oxidation.

7.3 Conclusions

Two optical probes were developed with the intention of investigating pure solid fuel combustion under oxy-fuel conditions and were successfully applied to the Technikum. This facility offers similar features as the Oxy-Fuel Burner and allows pure TBM combustion in swirled, turbulent flows.

Three operating conditions in air atmosphere were investigated. While an early combustion with a high reaction progress was evident in the regular and a locally lean operating condition, a locally rich flame was elongated and led to high CO concentrations. The agreement to chemical equilibrium calculations was found to range from fair to good for the most downstream measurement positions. Additionally, two oxy-fuel flames with different O_2-content in the combustion atmosphere were investigated and compared to the respective air flames. The combustion temperatures of both flames were lower than during air-blown operation. While the flame under Oxy25 atmosphere showed a similar behavior as the locally lean air-blown flame, the combustion of TBM under Oxy21 atmosphere was poor and similar to the locally rich flame under air atmosphere.

Presumably due to a doubled CO concentration in the volatiles and a higher volatile content, the CO concentrations generated by a combustion of TBM exceed those of RL. The temperature and water vapor profiles, however, were similar, which makes this regenerative fuel a promising alternative to solid fuels.

8 Summary and Outlook

In the coming decades, the combustion of gaseous and solid fossil fuels will globally remain a major source of primary energy. As conventional combustion processes of these fuels contribute to the global warming through emissions of large CO_2 quantities, the development of alternative combustion strategies such as oxy-fuel combustion is imperative.

This work aims for a hierarchical investigation of oxy-fuel combustion in a wide range of applications including laminar and turbulent, gaseous and solid fuel as well as non-premixed and premixed combustion. Similarities among the technical realizations allow conclusions regarding relevant processes and dominant differences between conventional air-blown and modern oxy-fuel combustion. The applications are influenced by a multitude of highly-coupled physical and chemical processes on a large variety of temporal and spatial scales, which impede experimental investigations and theoretical modeling of oxy-fuel combustion.

At the beginning of this work, the most relevant processes were described and an overview of individual models were given. In combustion applications, the mass, momentum, enthalpy and atomic mass-fractions are conserved but require transport equations, which were listed later on. As reactive flows can either be laminar or turbulent, basic concepts and characteristics of these regimes were introduced. Then, the fundamentals of gaseous and solid fuel combustion chemistry were outlined together with the properties of solid fuels. An introduction into the most relevant chemical, thermodynamic and physical changes accompanying a transition from air-blown to oxy-fuel combustion was given.

As conventional tactile or ex-situ measurement techniques affect the sensitive oxy-fuel processes, an introduction into non-intrusive molecular spectroscopy was given later on with a focus on absorption-spectroscopy-relevant optical properties and descriptions. The fundamentals were finalized by an introduction into the concept of residence time distributions their state-of-the-art measurement techniques for technical combustion systems.

To investigate oxy-fuel processes, a multi-parameter optical measurement system based on tunable diode laser spectroscopy was developed in this work which allows quantitative measurements of the concentrations and temperatures of multiple combustion-relevant species. Namely, transitions of H_2O-, CO_2-, O_2-, CH_4-, CO-, OH-, HCl- and C_2H_2-molecules were probed using nine lasers of different laser technologies. Strategies to quasi-simultaneously operate multiple of these lasers at one position were developed and different application-dependent optical setups were adapted. A careful analysis of the measurement uncertainty performed for each application separately ensured highly reliable datasets. As the measurement system relies on high-quality absorption-line coefficients which are not fully available in databases, gaps in these parameters were filled through either a review of relevant literature or own measurements in two specifically designed gas cells. The collisional broadening and pressure shift of five absorption transitions caused by combustion-relevant species were measured. These experiments led to a unique dataset of broadening and shift coefficients that carefully matched the few available reference data. In most cases, these coefficients were reported for the first time or their uncertainties significantly reduced. Parts of this measurement system were validated at a well-known air-blown flame with good agreement to reference data.

The developed measurement system allowed to determine the thermochemical state of a 2-D laminar, non-premixed oxy-fuel/methane flame at various locations. Differences to air-blown combustion were described. Due to a higher heat capacity, the adiabatic flame temperature under oxy-fuel combustion atmosphere is reduced. Under stoichiometric conditions, the experiments showed that this effect is intensified by an increased radiative heat transfer. The water vapor concentrations in a flame of higher oxygen content than air were measured to be increased. Due to a higher carbon content in the reactants and a therefore boosted water-gas-shift reaction, the CO concentrations were measured to be significantly higher in the oxy-fuel flame, with peak values of 11 %$_{Vol.}$. These measurements may serve as a validation reference for future numerical simulations of generic setups.

The turbulent, swirled oxy-fuel/methane combustion was investigated with respect to the gas residence time distribution and thermochemical state in a larger system. The gas residence times were studied using a newly developed measurement strategy which relied on pulse-shaped HCl-injections into the burner's primary flow. Subsequent measurements of the species concentration at the combustion chamber inlet and simultaneously at the outlet allowed to calculate the residence time distribution. Even though the necessary experimental effort for this strategy is increased, it was shown to be superior to other injection and measurement strategies. Measurements with this technique demonstrated that due to thermal expansion the mean residence time and the dispersion decreased when switching from non-reacting to reacting conditions. A lower flow velocity at the inflow caused a larger mean residence time and dispersion for the oxy-fuel operating conditions compared to air. The two investigated oxy-fuel conditions, however, were almost indistinguishable. A comparison to simple analytical models proved that the flow can be described by a series of a well-stirred reactor in combination with a plug flow with good agreement in terms of the residence time distributions.

The thermochemical measurements at the same system showed that the gas-flow of the near-burner flame and the tertiary flow were not well-mixed upstream of the combustion chamber outlet plate. Good mixing was only achieved at measurement positions just above this plate. Due to the globally lean operating conditions, the flue-gas temperatures were comparably low. The discrepancy of the experimentally measured temperature to an adiabatic equilibrium temperature calculation was higher for oxy-fuel operating conditions than for combustion under air atmosphere, probably due to an increased radiative heat transfer by the higher CO_2 concentrations and a longer residence time in the combustor and therefore diffusive heat transfer through the walls. The measured H_2O, CO_2 and O_2 concentrations in the experiment closely matched the results of the chemical equilibrium calculations. The CO and CH_4 concentrations, however, significantly differed from the equilibrium. This feature was attributed to a severe flame-quenching of the combustion by the cold tertiary flow and a therefore incomplete reaction progress and methane burnout rate. While this effect was evident for combustion under air and Oxy25 combustion atmosphere, the effect seemed to vanish in Oxy30 combustion, probably due to a higher oxygen-availability.

The heterogeneous, gas-assisted solid fuel combustion was investigated at the same system as the homogeneous combustion. It was shown that the previously developed measurement system was applicable to highly particle-laden flows, but to struggle from particle-adhesion at the walls and short residence times, as the latter may be influenced by the injection shape. The measured thermochemical state was compared to equilibrium calculations which showed that under most operating conditions the volatile burnout was high and indistinguishably from a complete burnout rate. The char burnout, however, was shown to be incomplete and depending on the fuel and the operating condition. Three different fuels were investigated as representatives of the most widely used solid fuels lignite and bituminous coal as well as torrefied biomass as a representative of alternative, regenerative fuels. While the char burnout rate of lignite and torrefied biomass were high, the bituminous coal showed a poor rate. This effect was attributed to the higher volatile content of the first two solid fuels. The quenching by the cold tertiary flow, which was evident under homogeneous combustion alone, was shown to be reduced by co-firing of solid fuels. Higher amounts of solid fuel, however, were shown to increase the CO concentrations, probably due to higher equilibrium concentrations.

A close-to-application facility allowed to study the oxy-fuel combustion of the regenerative fuel with a complete char burnout. Two optical probes were specifically designed for this facility and successfully implemented. The measurement of air-blown combustion of torrefied biomass for three different near-burner equivalence ratios showed that the flame is located close to the burner for the regular and a lean operating condition. For these conditions, the temperature as well as the H_2O and CO_2 concentrations were shown to be almost unchanged between the most upstream and further downstream measurement positions. The CO concentrations were reducing downstream due to a post-flame oxidation and were generally lower. At the most downstream position, they were close to equilibrium concentrations for the lean flame. The locally rich flame was shown to have an elongated flame and thus gradual increase of temperature as well as H_2O and CO_2 concentrations. The CO concentrations, however, were significantly increased and only slowly declined further downstream. The equilibrium water vapor concentrations carefully matched the experimentally measured values at the most downstream measurement position.

A comparison between air and two oxy-fuel operating conditions showed that a combustion under Oxy25 seems to be similar to the locally lean air-blown operation. The Oxy21 operating conditions, however, showed similar features as the locally rich air-blown flame, with locally high CO concentrations and a gradual increase of temperature and H_2O concentrations. Compared to a combustion of the conventional solid fuel lignite, the combustion of torrefied biomass showed higher CO concentrations probably due to a higher CO content in the volatile gases of this fuel. As the temperature and water vapor profiles, however, are similar, this fuel seems to be a suitable alternative to fossil fuels.

Summarizing this work, the following achievements were made. A measurement system for investigating the thermochemical state of oxy-fuel combustion systems was developed. This system allowed an experimental comparison of oxy-fuel and air-blown combustion in four combustion modes. Gas combustion, solid fuel combustion as well as a combination of both were investigated. A comparison to equilibrium calculations and analytical models revealed specific features of the chosen operating conditions, such as an incomplete burnout of the gaseous fuel or the char as well as elongated flames that let to unnecessarily high CO concentrations.

8.1 Outlook

For future investigations, the developed measurement system can be applied to other oxy-fuel combustion systems. As an example, experiments on an oxy-fuel counter-flow burner at the Institute for Technical Combustion, RWTH Aachen are scheduled and partially conducted already. For a higher flexibility and extended studies, the system could be expanded by lasers for NO, NO_2 and SO_3 which feature neat absorption lines in the MIR. Additional measurements of the temperature-dependent collisional broadening and the OH-absorption line strength [203] should allow to increase the accuracy of the measurement system.

In terms of combustion research, a comparison between the data measured at the WHP-burner and numerical simulations should provide helpful insights. Studies regarding this topic are currently under preparation. For the Oxy-Fuel Burner, a definition and investigation of more realistic operating conditions closer to stoichiometry are imperative and currently under development. A utilization of multi-pass optical setups in the Technikum should allow to detect minor species and as such provide deeper insights. Tomographic techniques or a combination of the system with imaging techniques should shed light on the three-dimensional features of this facility.

Appendix

Flue Gas Temperatures of Chapter 6

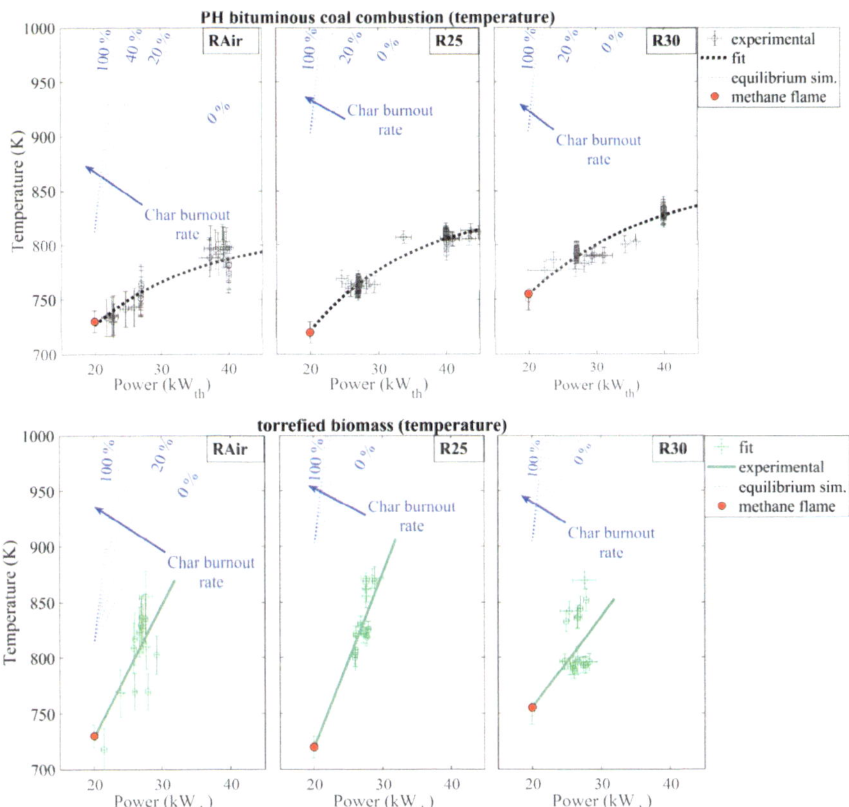

Fig. A.1: Flue gas temperatures of PHBM (top)/ TBM (bottom) – combustion in the stabilizing methane flame under RAir, R25 and R30. The points were measured with the two H_2O-spectrometers and, under oxy-fuel atmospheres, with the CO_2-spectrometer. As the coal feeder is not capable of constantly delivering 20 kW_{th} of TBM, a linear function is fitted to the TBM data. The bimodal distribution of temperatures in the R30/TBM operating condition probably stems from not having reached thermodynamic steady state conditions in the combustion chamber during one of the measurement.

Water Vapor Concentrations of Chapter 6

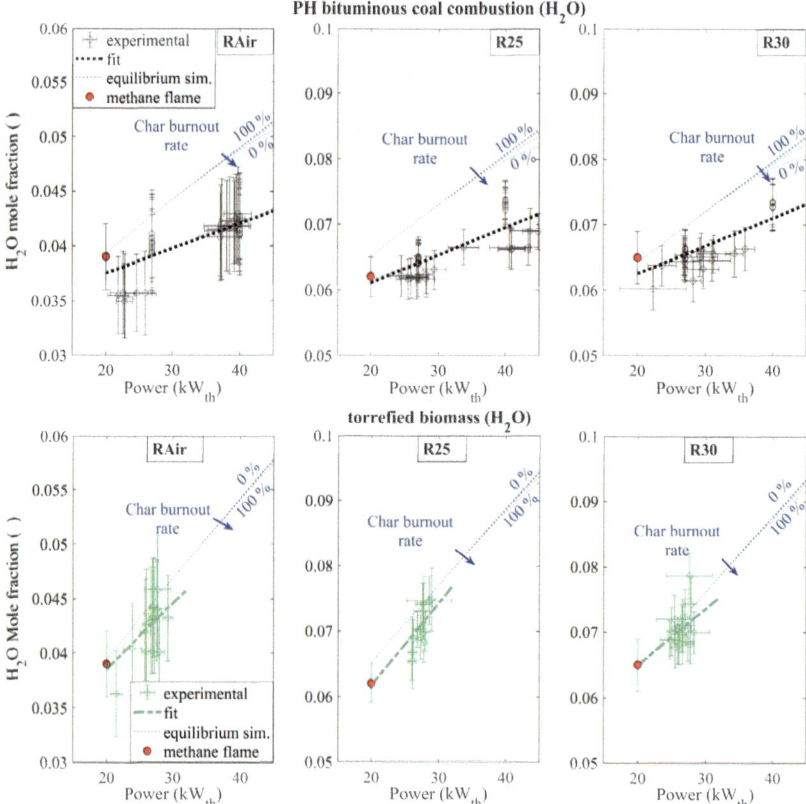

Fig. A.2: Water vapor mole fractions of gas-assisted PHBM (top)/ TBM (bottom) combustion for RAir, R25 and R30.

Carbon Dioxide Concentrations of Chapter 6

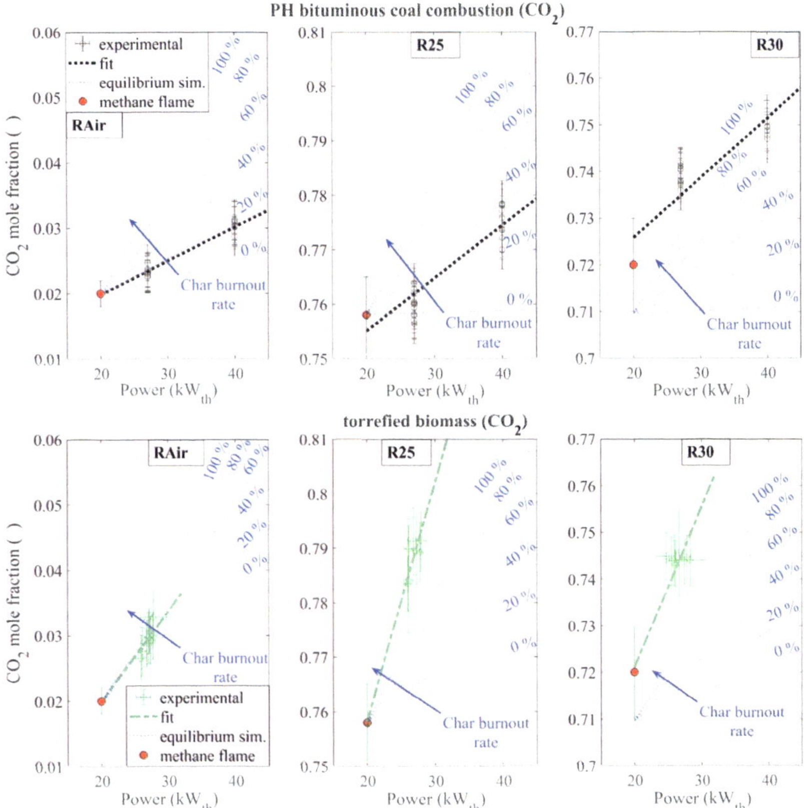

Fig. A.3: Carbon dioxide mole fractions of gas-assisted PHBM (top)/ TBM (bottom) combustion for RAir, R25 and R30.

Carbon Monoxide Concentrations of Chapter 6

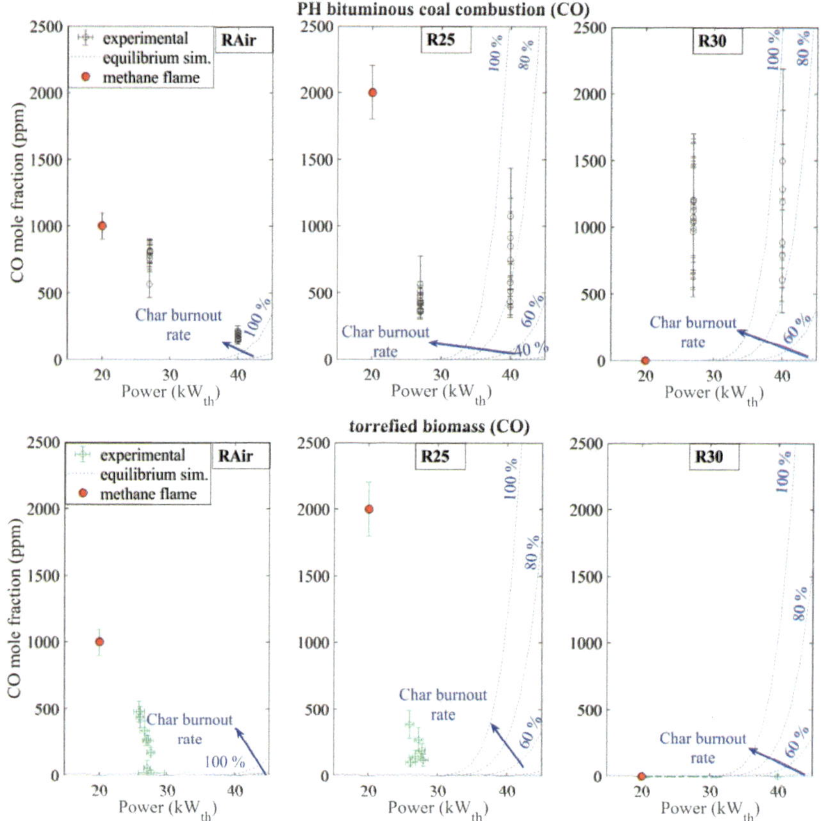

Fig. A.4: Carbon monoxide mole fractions of of PHBM (top)/ TBM (bottom) – combustion in the stabilizing methane flame under RAir, R25 and R30.

Gas Temperatures in the Technikum as mentioned in Chapter 7

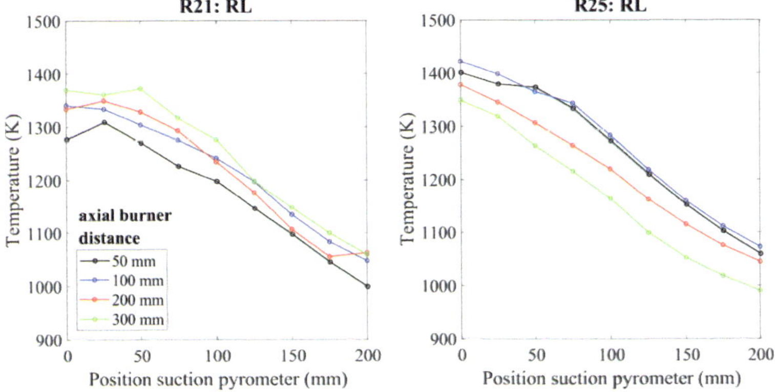

Fig. A.5: Gas temperatures of R21 and R25 in the combustion of RL, measured with a suction pyrometer.[44]

---

[44] Kindly provided by Institute of Heat and Mass Transfer, RWTH Aachen.

Statement on Own Publications

This work arose from research projects I conducted during my time as research associate at the institute of Reactive Flows and Diagnostics at the TU Darmstadt. The AG Turbo 2020 project 2.3.4A "Fortgeschrittene, laserbasierte In-situ Diagnostik in Brennerkammern", which was funded by the German Bundesministerium für Wirtschaft und Energie, enabled me to develop first spectrometers and get into contact with combustion research. The Collaborative Research Center/Transregio 129 "Oxy-Flame: Development of methods and models to describe solid fuel reactions within an oxy-fuel atmosphere", funded by Deutsche Forschungsgemeinschaft, allowed me to intensify the spectrometer development and to put a stronger focus on oxy-fuel combustion research. The indispensable support of both sponsors is kindly acknowledged.

Parts of this work have already been published in collaboration with other researchers. In particular, the publications [E1,E2,E5,E6,E10] and [E11] had the highest impact. Here, I would like to take the chance to thank all the co-authors of my publications again for their valuable contributions. However, dissertations these days are unfortunately not only required the show the capability of the author to perform sophisticated research, but also prove the independence and freedom of plagiarism. Hence, this general statement is intended to describe and separate the contributions of the co-authors. In addition, it is necessary to mention that I was listed as primary author of these publications to emphasize the importance of my contributions.

The experimental section of the line data measurements (section 3.5.2 and 3.5.3) was published in [E2]. Here, my student Nicole Walter conducted most of the experiments and supported me with valuable information on the literature. The general idea of the experiments was developed together with my advisor Steven Wagner. My duties were to plan the experiments in detail, develop and construct the experimental cells, the post-processing and analysis of the results.

Steven Wagner enabled me to conduct experiments on the version of the Wolfhard-Parker burner he developed and constructed, which led to the results in chapter 4 and the publication [E5]. Both Steven Wagner and Volker Ebert provided assistance in the post-processing of the data through valuable information on the general post-processing strategy and provision of the in-house post-processing code. Andreas Dreizler helped with the interpretation of the results in terms of questions about flames and combustion processes. The details of the experimental setup, in particular the spectrometers, were designed and constructed by myself. In addition, I performed the experiments and major parts of the post-processing and interpretation.

The Oxy-Flame Burner, which was essential for the experiments in chapters 5 and 6, was designed, constructed and partially operated by Lukas Georg Becker, who also defined the operating conditions. The experiments let to results published in [E6] and [E10]. In both publications, Andreas Dreizler and Steven Wagner supervised the experiments and supported the analysis of the results in terms of combustion-related and spectroscopic interpretations, respectively. Maria Angela Agizza provided background knowledge on chemical reactor networks and residence time distributions.

Even though the residence time measurements and the necessary technique presented in this thesis (sections 5.2 and 6.2) were entirely new developments, they are based on measurements of the cumulative distribution function published in [E1]. Here, Maria Angela Agizza supported me with the valuable reactor models and Lukas Georg Becker enabled me to perform the investigations on the combustion chamber he developed and operated. Steven Wagner, Volker Ebert and Andreas Dreizler supervised the investigations and supported the interpretation process of the results.

The measurements of chapter 7 are based on two measurement campaigns I have been allowed to perform at the WSA-Technikum Aachen. Some results of these campaigns are published in [E11]. Here,

Dominik Rauen developed the probes which granted optical access for the TDLAS measurements into the combustion chamber. Johannes Hees, Diego Zabrodiec and Anna Maßmeyer operated this burner and assisted the measurements. Georg Möller supplied us with torrefied biomass he developed and produced. Volker Ebert, Reinhold Kneer, Andreas Dreizler and Steven Wagner supervised the experiments, enabled the collaborations between Aachen and Darmstadt and helped with the interpretation of the results.

Bibliography

[1] Aiello LC, Wheeler P. The Expensive-Tissue Hypothesis: The Brain and the Digestive System in Human and Primate Evolution. Current Anthropology 1995;36(2):199–221.

[2] World energy outlook. Paris: OECD/IEA; 2016.

[3] Intergovernmental Panel on Climate Change. Climate Change 2014: Synthesis Report. Contribution of Working Groups I, II and III to the Fifth Assessment Report of the Intergovernmental Panel on Climate Change. IPCC, Geneva, Switzerland 2014;2014:151 pp.

[4] United Nations Organization. Paris Agreement; 2015.

[5] Intergovernmental Panel on Climate Change. Carbon Capture and Storage. Technical report; 2005.

[6] Technical Report. The Global Status of CCS: 2017; 2017.

[7] Toftegaard MB, Brix J, Jensen PA, Glarborg P, Jensen AD. Oxy-fuel combustion of solid fuels. Progress in Energy and Combustion Science 2010;36(5):581–625.

[8] Abraham BM, Asbury JG, Lynch EP, Teotia APS. Coal-Oxygen Process Provides CO_2 for Enhanced Recovery. Oil and Gas Journal;1982(80 (11)).

[9] Hu Y, Naito S, Kobayashi N, Hasatani M. CO_2, NO_x and SO_2 emissions from the combustion of coal with high oxygen concentration gases. Fuel 2000;79(15):1925–32.

[10] Ditaranto M, Oppelt T. Radiative heat flux characteristics of methane flames in oxy-fuel atmospheres. Experimental Thermal and Fluid Science 2011;35(7):1343–50.

[11] Hu X, Yu Q, Liu J, Sun N. Investigation of laminar flame speeds of $CH_4/O_2/CO_2$ mixtures at ordinary pressure and kinetic simulation. Energy 2014;70:626–34.

[12] Xie Y, Wang J, Zhang M, Gong J, Jin W, Huang Z. Experimental and Numerical Study on Laminar Flame Characteristics of Methane Oxy-fuel Mixtures Highly Diluted with CO_2. Energy Fuels 2013;27(10):6231–7.

[13] Zhu DL, Egolfopoulos FN, Law CK. Experimental and numerical determination of laminar flame speeds of methane/(Ar, N_2, CO_2)-air mixtures as function of stoichiometry, pressure, and flame temperature. Symposium (International) on Combustion 1989;22(1):1537–45.

[14] Clarke A, Hargrave GK. Measurements of laminar premixed methane—air flame thickness at ambient conditions. Proceedings of the Institution of Mechanical Engineers, Part C: Journal of Mechanical Engineering Science 2009;223(8):1969–73.

[15] Hainsworth D, Pourkashanian M, Richardson AP, Rupp JL, Williams A. The influence of carbon dioxide on smoke formation and stability in methane-oxygen-carbon dioxide flames. Fuel 1996;75(3):393–6.

[16] Ditaranto M, Hals J. Combustion instabilities in sudden expansion oxy–fuel flames. Combustion and Flame 2006;146(3):493–512.

[17] Sevault A. Investigation of the structure of oxy-fuel flames using Raman laser diagnostics [Dissertation]. Tronheim, Norway: Norwegian University of Science and Technology.

[18] Kutne P, Kapadia BK, Meier W, Aigner M. Experimental analysis of the combustion behaviour of oxyfuel flames in a gas turbine model combustor. Proceedings of the Combustion Institute 2011;33(2):3383–90.

[19] Li G-n, Zhou H, Cen K-f. Emission characteristics and combustion instabilities in an oxy-fuel swirl-stabilized combustor. J. Zhejiang Univ. Sci. A 2008;9(11):1582–9.

[20] Kapadia BK. Experimental Study of Oxyfuel Combustion for Stationary Gas Turbine Applications [Dissertation]. Stuttgart: University of Stuttgart; 2016.

[21] Ardha VR, Dam B, Love N, Choudhuri A. Characterization of Oxy-Fuel flames in a Swirl Based Combustor. 10th International Energy Conversion Engineering Conference;2012(AIAA 2012-3786).

[22] Sevault A, Dunn M, Barlow RS, Ditaranto M. On the structure of the near field of oxy-fuel jet flames using Raman/Rayleigh laser diagnostics. Combustion and Flame 2012;159(11):3342–52.

[23] Becker LG, Kosaka H, Böhm B, Doost S, Knappstein R, Habermehl M et al. Experimental investigation of flame stabilization inside the quarl of an oxyfuel swirl burner. Fuel 2017;201:124–35.

[24] Kim HK, Kim Y, Lee SM, Ahn KY. NO reduction in 0.03–0.2MW oxy-fuel combustor using flue gas recirculation technology. Proceedings of the Combustion Institute 2007;31(2):3377–84.

[25] Becher V, Goanta A, Gleis S, Spliethoff H. Controlled Staging with Non-Stoichiometric Burners for Oxyfuel Processes. 32nd International Technical Conference on Coal Utilization & Fuel Systems 2007(Clearwater Florida).

[26] Tan Y, Douglas MA, Thambimuthu KV. CO_2 capture using oxygen enhanced combustion strategies for natural gas power plants. Fuel 2002;81(8):1007–16.

[27] Mathieu P, Nihart R. Zero-Emission MATIANT Cycle. J. Eng. Gas Turbines Power 1999;121(1):116.

[28] Yantovski EI, Zvagolsky KN, Gavrilenko VA. The COOPERATE-demo power cycle. Energy Conversion and Management 1995;36(6-9):861–4.

[29] Bolland O, Mathieu P. Comparison of two CO_2 removal options in combined cycle power plants. Energy Conversion and Management 1998;39(16-18):1653–63.

[30] Bolland O, Sæther S. New concepts for natural gas fired power plants which simplify the recovery of carbon dioxide. Energy Conversion and Management 1992;33(5-8):467–75.

[31] Hees J, Zabrodiec D, Massmeyer A, Pielsticker S, Gövert B, Habermehl M et al. Detailed analyzes of pulverized coal swirl flames in oxy-fuel atmospheres. Combustion and Flame 2016;172:289–301.

[32] Hees J, Zabrodiec D, Massmeyer A, Habermehl M, Kneer R. Experimental Investigation and Comparison of Pulverized Coal Combustion in CO_2/O_2- and N_2/O_2-Atmospheres. Flow Turbulence Combust 2016;96(2):417–31.

[33] Köser J, Becker LG, Goßmann A-K, Böhm B, Dreizler A. Investigation of ignition and volatile combustion of single coal particles within oxygen-enriched atmospheres using high-speed OH-PLIF. Proceedings of the Combustion Institute 2017;36(2):2103–11.

[34] Jia L, Tan Y, McCalden D, Wu Y, He I, Symonds R et al. Commissioning of a 0.8MWth CFBC for oxy-fuel combustion. International Journal of Greenhouse Gas Control 2012;7:240–3.

[35] Anheden M, Burchhardt U, Ecke H, Faber R, Jidinger O, Giering R et al. Overview of operational experience and results from test activities in Vattenfall's 30 MWth oxyfuel pilot plant in Schwarze Pumpe 2011;4:941–50.

[36] PAYNE R, Chen LS, Wolksy AM, RICHTER WF. CO 2 Recovery via Coal Combustion in Mixtures of Oxygen and Recycled Flue Gas. Combust. Sci. Technol. 1989;67(1-3):1–16.

[37] Balusamy S, Kamal MM, Lowe SM, Tian B, Gao Y, Hochgreb S. Laser diagnostics of pulverized coal combustion in O_2/N_2 and O_2/CO_2 conditions: Velocity and scalar field measurements. Exp Fluids 2015;56(5):1.

[38] Balusamy S, Schmidt A, Hochgreb S. Flow field measurements of pulverized coal combustion using optical diagnostic techniques. Exp Fluids 2013;54(5):159.

[39] Molina A, Shaddix CR. Ignition and devolatilization of pulverized bituminous coal particles during oxygen/carbon dioxide coal combustion. Proceedings of the Combustion Institute 2007;31(2):1905–12.

[40] Shaddix CR, Molina A. Particle imaging of ignition and devolatilization of pulverized coal during oxy-fuel combustion. Proceedings of the Combustion Institute 2009;32(2):2091–8.

[41] Bejarano PA, Levendis YA. Single-coal-particle combustion in O_2/N_2 and O_2/CO_2 environments. Combustion and Flame 2008;153(1-2):270–87.

[42] Kim Y-G, Kim J-D, Lee B-H, Song J-H, Chang Y-J, Jeon C-H. Experimental Investigation into Combustion Characteristics of Two Sub-bituminous Coals in O_2/N_2 and O_2/CO_2 Environments. Energy Fuels 2010;24(11):6034–40.

[43] Bai X, Lu G, Bennet T, Sarroza A, Eastwick C, Liu H et al. Combustion behavior profiling of single pulverized coal particles in a drop tube furnace through high-speed imaging and image analysis. Experimental Thermal and Fluid Science 2017;85:322–30.

[44] Riaza J, Gibbins J, Chalmers H. Ignition and combustion of single particles of coal and biomass. Fuel 2017;202:650–5.

[45] Lee H, Choi S. An observation of combustion behavior of a single coal particle entrained into hot gas flow. Combustion and Flame 2015;162(6):2610–20.

[46] Vorobiev N, Geier M, Schiemann M, Scherer V. Experimentation for char combustion kinetics measurements: Bias from char preparation. Fuel Processing Technology 2016;151:155–65.

[47] Schiemann M, Geier M, Shaddix CR, Vorobiev N, Scherer V. Determination of char combustion kinetics parameters: Comparison of point detector and imaging-based particle-sizing pyrometry. Rev Sci Instrum 2014;85(7):75114.

[48] Wang S, Lu H, Zhao Y, Mostofi R, Young Kim H, Yin L. Numerical study of coal particle cluster combustion under quiescent conditions. Chemical Engineering Science 2007;62(16):4336–47.

[49] Smith NL, Nathan GJ, Zhang DK, Nobes DS. The significance of particle clustering in pulverized coal flames. Proceedings of the Combustion Institute 2002;29(1):797–804.

[50] Liu H, Zheng S, Zhou H. Measurement of Soot Temperature and Volume Fraction of Axisymmetric Ethylene Laminar Flames Using Hyperspectral Tomography. IEEE Trans. Instrum. Meas. 2017;66(2):315–24.

[51] Dhungel B. Experimental Investigations on Combustion and Emission Behaviour during oxy-Coal Combustion [Dissertation]. Stuttgart: Universität Stuttgart.

[52] Drasek WA von, Charon O, Mulderink K, Sonnenfroh DM, Allen MG. Multifunctional industrial combustion process monitoring with tunable diode lasers. In: Farquharson S, editor. Environmental and Industrial Sensing. SPIE; 2001, p. 133–141.

[53] Zabrodiec D, Becker L, Hees J, Maßmeyer A, Habermehl M, Hatzfeld O et al. Detailed Analysis of the Velocity Fields from 60 kW Swirl-Stabilized Coal Flames in CO 2 /O 2 - and N 2 /O 2 - Atmospheres by Means of Laser Doppler Velocimetry and Particle Image Velocimetry. Combustion Science and Technology 2017;189(10):1751–75.

[54] Habermehl M, Hees J, Maßmeyer A, Zabrodiec D, Hatzfeld O, Kneer R. Comparison of Flame Stability Under Air and Oxy-Fuel Conditions for an Aerodynamically Stabilized Pulverized Coal Swirl Flame. J. Energy Resour. Technol 2016;138(4):42209.

[55] Scheffknecht G, Al-Makhadmeh L, Schnell U, Maier J. Oxy-fuel coal combustion—A review of the current state-of-the-art. International Journal of Greenhouse Gas Control 2011;5:S16-S35.

[56] Navier, C. L. M. H. Mémoire sur les lois du Mouvement des Fluides. Mémoires de l'Académie Royale des Sciences de l'Institut de France;1823:389–440.

[57] Stokes GG. On the theories of the internal friction of fluids in motion. Trans. Cambridge Philos. Soc. 1845(8):287–319.

[58] Reynolds O. An experimental investigation of the circumstances which determine whether the motion of water shall be direct or sinuous and the law of resistance in parallel channels. Phil. Trans. R. Soc;1883(174):935–82.

[59] Richardson LF. The Supply of Energy from and to Atmospheric Eddies. Proceedings of the Royal Society A: Mathematical, Physical and Engineering Sciences 1920;97(686):354–73.

[60] Kolmogorov A. The Local Structure of Turbulence in Incompressible Viscous Fluid for Very Large Reynolds' Numbers. Doklady Akademiia Nauk SSSR 1941(30):301–5.

[61] Smith GP, Golden DM, Frenklach M, Moriarty MW, Eiteneer B, Goldenberg M et al. GRI-Mech 3.0; Available from: http://www.me.berkeley.edu/gri_mech/.

[62] Liu Y, Sun X, Sethi V, Nalianda D, Li Y-G, Wang L. Review of modern low emissions combustion technologies for aero gas turbine engines. Progress in Aerospace Sciences 2017;94:12–45.

[63] Bilger RW, Stårner SH, Kee RJ. On reduced mechanisms for methane · air combustion in nonpremixed flames. Combustion and Flame 1990;80(2):135–49.

[64] Vagelopoulos CM, Egolfopoulos FN. Direct experimental determination of laminar flame speeds. Symposium (International) on Combustion 1998;27(1):513–9.

[65] Warnatz J, Maas U, Dibble R (eds.). Combustion. Springer Berlin Heidelberg; 2006.

[66] Peters N. Turbulent Combustion. Cambridge: Cambridge University Press; 2000.

[67] Joos F. Technische Verbrennung: Verbrennungstechnik, Verbrennungsmodellierung, Emissionen mit 65 Tabellen. Berlin, Heidelberg: Springer-Verlag Berlin Heidelberg; 2006.

[68] Knappstein R. Large Eddy Simulation of Pulverized Coal Combustion in the Context of Tabulated Chemistry. 1st ed. Aachen: Shaker; 2018.

[69] Vassilev SV, Baxter D, Andersen LK, Vassileva CG. An overview of the chemical composition of biomass. Fuel 2010;89(5):913–33.

[70] Chew JJ, Doshi V. Recent advances in biomass pretreatment – Torrefaction fundamentals and technology. Renewable and Sustainable Energy Reviews 2011;15(8):4212–22.

[71] Zelkowski J. Kohlecharakterisierung und Kohleverbrennung - Kohle als Brennstoff, Physik und Theorie der Kohleverbrennung. Essen, Germany: VGB PowerTech Service GmbH; 2004.

[72] Smoot LD, Smith PJ. Coal Combustion and Gasification. Boston, MA, s.l.: Springer US; 1985.

[73] Spliethoff H. Power Generation from Solid Fuels. 1st ed. s.l.: Springer-Verlag; 2010.

[74] Smith KL, Smoot LD, Fletcher TH, Pugmire RJ. The Structure and Reaction Processes of Coal. Boston, MA: Springer; 1994.

[75] Buhre BJP, Elliott LK, Sheng CD, Gupta RP, Wall TF. Oxy-fuel combustion technology for coal-fired power generation. Progress in Energy and Combustion Science 2005;31(4):283–307.

[76] Goettlicher G. Energetik der Kohlendioxidrueckhaltung in Kraftwerken: Energetics of carbon dioxide retention in power plants. Fortschritt-Berichte VDI, Reihe 6: Energietechnik, VDI Verlag;1999(421):1–187.

[77] Joseph Fraunhofer. Bestimmung des Brechungs- und des Farben-Zerstreuungs – Vermögens verschiedener Glasarten, in Bezug auf die Vervollkommnung achromatischer Fernröhre. Annalen der Physik;1817(56):264–313.

[78] Planck M. Über irreversible Strahlungsvorgänge. Sitzungsberichte der Königlich Preußischen Akademie der Wissenschaften zu Berlin 1899;1899(Erster Halbband):479–80.

[79] Einstein A. Über einen die Erzeugung und Verwandlung des Lichtes betreffenden heuristischen Gesichtspunkt. Ann. Phys. Chem. 1905;322(6):132–48.

[80] Bohr N. I. On the constitution of atoms and molecules. The London, Edinburgh, and Dublin Philosophical Magazine and Journal of Science 2009;26(151):1–25.

[81] Broglie L-V de. Recherches sur la théorie des quanta [Dissertation]. Paris: Masson et Cie; 1924.

[82] Heisenberg W. ber den anschaulichen Inhalt der quantentheoretischen Kinematik und Mechanik. Z. Physik 1927;43(3-4):172–98.

[83] Demtröder W. Atoms, Molecules and Photons: An Introduction to Atomic-, Molecular- and Quantum Physics. 2nd ed. Berlin, Heidelberg: Springer-Verlag Berlin Heidelberg; 2010.

[84] Atkins PW, Friedman R. Molecular quantum mechanics. 5th ed. Oxford: Oxford Univ. Press; 2011.

[85] Reinhold J. Quantentheorie der Moleküle: Eine Einführung. 5th ed. Wiesbaden: Springer Spektrum; 2015.

[86] Mayinger F, Feldmann O (eds.). Optical Measurements: Techniques and Applications. Berlin, Heidelberg: Springer; 2001.

[87] Demtröder W. Laserspektroskopie. Berlin, Heidelberg: Springer Berlin Heidelberg; 2007.

[88] Gordon IE, Rothman LS, Hill C, Kochanov RV, Tan Y, Bernath PF et al. The HITRAN2016 molecular spectroscopic database. Journal of Quantitative Spectroscopy and Radiative Transfer 2017;203:3–69.

[89] Rothman LS, Gordon IE, Barber RJ, Dothe H, Gamache RR, Goldman A et al. HITEMP, the high-temperature molecular spectroscopic database. Journal of Quantitative Spectroscopy and Radiative Transfer 2010;111(15):2139–50.

[90] Sobel'man II, Sobel'man II, Vainshtein LA, Yukov EA, Ukov EA. Excitation of Atoms and Broadening of Spectral Lines. Berlin, New York: Springer; 1995.

[91] Dicke RH. The Effect of Collisions upon the Doppler Width of Spectral Lines. Phys. Rev. 1953;89(2):472–3.

[92] Pierluissi JH, Vanderwood PC, Gomez RB. Fast calculational algorithm for the Voigt profile. Journal of Quantitative Spectroscopy and Radiative Transfer 1977;18(5):555–8.

[93] Humlíček J. An efficient method for evaluation of the complex probability function: The Voigt function and its derivatives. Journal of Quantitative Spectroscopy and Radiative Transfer 1979;21(4):309–13.

[94] Abramowitz M, Stegun IA (eds.). Handbook of mathematical functions: With formulas, graphs, and mathematical tables. 9th ed. New York, NY: Dover Publ; 2013.

[95] Tennyson J, Bernath PF, Campargue A, Császár AG, Daumont L, Gamache RR et al. Recommended isolated-line profile for representing high-resolution spectroscopic transitions (IUPAC Technical Report). Pure and Applied Chemistry 2014;86(12):1931–43.

[96] Goldenstein CS, Spearrin RM, Jeffries JB, Hanson RK. Infrared laser-absorption sensing for combustion gases. Progress in Energy and Combustion Science 2017;60:132–76.

[97] Wagner S, Fisher BT, Fleming JW, Ebert V. TDLAS-based in situ measurement of absolute acetylene concentrations in laminar 2D diffusion flames. Proceedings of the Combustion Institute 2009;32(1):839–46.

[98] Webber ME, Wang J, Sanders ST, Baer DS, Hanson RK. In situ combustion measurements of CO, CO_2, H_2O and temperature using diode laser absorption sensors. Proceedings of the Combustion Institute 2000;28(1):407–13.

[99] Teichert H. Entwicklung und Einsatz von Diodenlaser-Spektrometern zur simultanen In-situ-Detektion von CO, O_2 und H_2O in technischen Verbrennungsprozessen [Heidelberg University Library]; 2003.

[100] Sun K, Sur R, Chao X, Jeffries JB, Hanson RK, Pummill RJ et al. TDL absorption sensors for gas temperature and concentrations in a high-pressure entrained-flow coal gasifier. Proceedings of the Combustion Institute 2013;34(2):3593–601.

[101] Mackay KL, Chanda A, Mackay G, Pisano JT, Durbin TD, Crabbe K et al. Measurement of Hydrogen Chloride in Coal-Fired Power Plant Emissions Using Tunable Diode Laser Spectrometry. J Appl Spectrosc 2016;83(4):627–33.

[102] Sur R, Sun K, Jeffries JB, Hanson RK, Pummill RJ, Waind T et al. TDLAS-based sensors for in situ measurement of syngas composition in a pressurized, oxygen-blown, entrained flow coal gasifier. Appl. Phys. B 2014;116(1):33–42.

[103] Stritzke F, van der Kley S, Feiling A, Dreizler A, Wagner S. Ammonia concentration distribution measurements in the exhaust of a heavy duty diesel engine based on limited data absorption tomography. Opt Express 2017;25(7):8180–91.

[104] Diemel O, Pareja J, Dreizler A, Wagner S. An Interband Cascade Laser-Based In Situ Absorption Sensor for Nitric Oxide in Combustion Exhaust Gases. Appl. Phys. B 2017;123(5):545.

[105] Witzel O, Klein A, Meffert C, Wagner S, Kaiser S, Schulz C et al. VCSEL-based, high-speed, in situ TDLAS for in-cylinder water vapor measurements in IC engines. Opt Express 2013;21(17):19951–65.

[106] Ortwein P, Woiwode W, Fleck S, Eberhard M, Kolb T, Wagner S et al. Absolute diode laser-based in situ detection of HCl in gasification processes. Exp Fluids 2010;49(4):961–8.

[107] Kranendonk LA, An X, Caswell AW, Herold RE, Sanders ST, Huber R et al. High speed engine gas thermometry by Fourier-domain mode-locked laser absorption spectroscopy. Opt. Express 2007;15(23):15115.

[108] Schultz IA, Goldenstein CS, Mitchell Spearrin R, Jeffries JB, Hanson RK, Rockwell RD et al. Multispecies Midinfrared Absorption Measurements in a Hydrocarbon-Fueled Scramjet Combustor. Journal of Propulsion and Power 2014;30(6):1595 604.

[109] Sanders ST, Wang J, Jeffries JB, Hanson RK. Diode-laser absorption sensor for line-of-sight gas temperature distributions. Appl. Opt. 2001;40(24):4404.

[110] Liu C, Xu L, Cao Z. Measurement of nonuniform temperature and concentration distributions by combining line-of-sight tunable diode laser absorption spectroscopy with regularization methods. Appl. Opt. 2013;52(20):4827–42.

[111] Wang J, Maiorov M, Baer DS, Garbuzov DZ, Connolly JC, Hanson RK. In situ combustion measurements of CO with diode-laser absorption near 23 µm. Appl. Opt. 2000;39(30):5579.

[112] Baer DS, Nagali V, Furlong ER, Hanson RK, Newfield ME. Scanned- and fixed-wavelength absorption diagnostics for combustion measurements using multiplexed diode lasers. AIAA Journal 1996;34(3):489–93.

[113] Rieker GB, Li H, Liu X, Liu JTC, Jeffries JB, Hanson RK et al. Rapid measurements of temperature and H_2O concentration in IC engines with a spark plug-mounted diode laser sensor. Proceedings of the Combustion Institute 2007;31(2):3041–9.

[114] Sepman A, Ögren Y, Qu Z, Wiinikka H, Schmidt FM. Real-time in situ multi-parameter TDLAS sensing in the reactor core of an entrained-flow biomass gasifier. Proceedings of the Combustion Institute 2017;36(3):4541–8.

[115] Teichert H, Fernholz T, Ebert V. Simultaneous in situ measurement of CO, H_2O, and gas temperatures in a full-sized coal-fired power plant by near-infrared diode lasers. Appl. Opt. 2003;42(12):2043.

[116] Rieker GB. Wavelength modulation spectroscopy for measurements of gas temperature and concentration in hars environments [Dissertation]. Standford: Stanford University; 2009.

[117] Goldenstein CS, Strand CL, Schultz IA, Sun K, Jeffries JB, Hanson RK. Fitting of calibration-free scanned-wavelength-modulation spectroscopy spectra for determination of gas properties and absorption lineshapes. Appl Opt 2014;53(3):356–67.

[118] Werle P, Mücke R, Slemr F. The limits of signal averaging in atmospheric trace-gas monitoring by tunable diode-laser absorption spectroscopy (TDLAS). Appl. Phys. B 1993;57(2):131–9.

[119] IEEE Standard Specification Format Guide and Test Procedure for Single-Axis Interferometric Fiber Optic Gyros. Piscataway, NJ, USA: IEEE; 1997. doi:10.1109/ieeestd.1998.86153.

[120] Werle P. Tunable diode laser absorption spectroscopy: Recent findings and novel approaches. Infrared Physics & Technology 1996;37(1):59–66.

[121] Bowling DR, Sargent SD, Tanner BD, Ehleringer JR. Tunable diode laser absorption spectroscopy for stable isotope studies of ecosystem–atmosphere CO_2 exchange. Agricultural and Forest Meteorology 2003;118(1-2):1–19.

[122] Fried A, Henry B, Wert B, Sewell S, Drummond JR. Laboratory, ground-based, and airborne tunable diode laser systems: Performance characteristics and applications in atmospheric studies. Applied Physics B: Lasers and Optics 1998;67(3):317–30.

[123] Phelan R, O'Carroll J, Byrne D, Herbert C, Somers J, Kelly B. InGaAs/InP Multiple Quantum-Well Discrete-Mode Laser Diode Emitting at 2 µm. IEEE Photon. Technol. Lett. 2012;24(8):652–4.

[124] Phelan R, Guo W-H, Lu Q, Byrne D, Roycroft B, Lambkin P et al. A Novel Two-Section Tunable Discrete Mode Fabry-PÉrot Laser Exhibiting Nanosecond Wavelength Switching. IEEE J. Quantum Electron. 2008;44(4):331–7.

[125] Anandarajah P, Perry P, Herbert C, Jones D, Kaszubowska-Anandarajah A, Barry LP et al. Discrete mode lasers for communication applications. IET Optoelectronics 2009;3(1):1–17.

[126] Numai T. Fundamentals of Semiconductor Lasers. 2nd ed. Tokyo: Springer; 2015.

[127] Eichler J, Eichler H-J. Laser: Bauformen, Strahlführung, Anwendungen. 7th ed. Berlin, Heidelberg: Springer-Verlag Berlin Heidelberg; 2010.

[128] Ghafouri–Shiraz H. Distributed Feedback Laser Diodes and Optical Tunable Filters. Chichester, UK: John Wiley & Sons, Ltd; 2003.

[129] Morthier G. Handbook of Distributed Feedback Laser Diodes. 2nd ed. Norwood: Artech House; 2013.

[130] Yu SF. Analysis and Design of Vertical Cavity Surface Emitting Lasers. Hoboken, NJ, USA: John Wiley & Sons, Inc; 2003.

[131] Michalzik R. VCSELs: Fundamentals, technology and applications of vertical-cavity surface-emitting lasers. Berlin: Springer; 2013.

[132] Kneubühl FK, Sigrist MW. Laser. 7th ed. Wiesbaden: Vieweg + Teubner; 2008.

[133] Göckeler K, Terhaar S, Oliver Paschereit C. Residence Time Distribution in a Swirling Flow at Nonreacting, Reacting, and Steam-Diluted Conditions. J. Eng. Gas Turbines Power 2014;136(4):41505.

[134] Pedersen LS, Breithauptb P, Johansen KD, Weber R. Residence Time Distributions in Confined Swirling Flames. Combustion Science and Technology 1997;127(1-6):251–73.

[135] Beér JM, Lee KB. The effect of the residence time distribution on the performance and efficiency of combustors. Symposium (International) on Combustion 1965;10(1):1187–202.

[136] Faravelli T, Bua L, Frassoldati A, Antifora A, Tognotti L, Ranzi E. A new procedure for predicting NOx emissions from furnaces. Computers & Chemical Engineering 2001;25(4-6):613–8.

[137] Cheng L, Spencer A. Residence time measurement of an isothermal combustor flow field. Exp Fluids 2012;52(3):647–61.

[138] Hagen J. Chemiereaktoren. Weinheim, FRG: Wiley-VCH Verlag GmbH & Co. KGaA; 2004.

[139] Nauman EB. Residence Time Distributions. In: Paul EL, Atiemo-Obeng VA, Kresta SM, editors. Handbook of Industrial Mixing. Hoboken, NJ, USA: John Wiley & Sons, Inc; 2003, p. 1–17.

[140] Weiß S (ed.). Verfahrenstechnische Berechnungsmethoden. 1st ed. Weinheim: VCH Verl.-Ges; 1987.

[141] Levenspiel O. Chemical reaction engineering. 3rd ed. Hoboken, NJ: Wiley; 1999.

[142] Dhungel B. Experimental investigations on combustion and emission behaviour during oxy-coal combustion. Universität Stuttgart; 2010.

[143] Goeckeler K, Terhaar S, Lacarelle A, Paschereit C. Residence Time Distribution in a Swirl-Stabilized Combustor at Cold Conditions. 41st AIAA Fluid Dynamics Conference and Exhibit 2014;2014.

[144] van der Lans RP, Glarborg P, Dam-Johansen K, Larsen PS. Residence time distributions in a cold, confined swirl flow. Chemical Engineering Science 1997;52(16):2743–56.

[145] Shin D, Park S, Jeon B, Yu T, Hwang J. Effect of swirling flow by normal injection of secondary air on the gas residence time and mixing characteristics in a lab-scale cold model combustor. J Mech Sci Technol 2006;20(12):2284–91.

[146] Rao JS, Ramani NVS, Pant JH, Reddy DH. Measurement of Residence Time Distributions Of Coal Particles in a Pressurized Fluidized Bed Gasifier (PFBG) using Radio Tracer Technique. Indian Journal of Science and Technology 2012;2012(5(12)):48–54.

[147] Schlosser E, Wolfrum J, Hildebrandt L, Seifert H, Oser B, Ebert V. Diode laser based in situ detection of alkali atoms: Development of a new method for determination of residence-time distribution in combustion plants. Appl Phys B 2002;75(2-3):237–47.

[148] Fuest F, Barlow RS, Chen J-Y, Dreizler A. Raman/Rayleigh scattering and CO-LIF measurements in laminar and turbulent jet flames of dimethyl ether. Combustion and Flame 2012;159(8):2533–62.

[149] Gregor MA, Seffrin F, Fuest F, Geyer D, Dreizler A. Multi-scalar measurements in a premixed swirl burner using 1D Raman/Rayleigh scattering. Proceedings of the Combustion Institute 2009;32(2):1739–46.

[150] Linow S, Dreizler A, Janicka J, Hassel EP. Measurement of temperature and concentration in oxy-fuel flames by Raman/Rayleigh spectroscopy. Meas. Sci. Technol. 2002;13(12):1952–61.

[151] Williams B, Edwards M, Stone R, Williams J, Ewart P. High precision in-cylinder gas thermometry using Laser Induced Gratings: Quantitative measurement of evaporative cooling with gasoline/alcohol blends in a GDI optical engine. Combustion and Flame 2014;161(1):270–9.

[152] Hasegawa R, Sakata I, Yanagihara H, Johansson B, Omrane A, Aldén M. Two-dimensional gas-phase temperature measurements using phosphor thermometry. Appl. Phys. B 2007;88(2):291–6.

[153] Abram C, Fond B, Heyes AL, Beyrau F. High-speed planar thermometry and velocimetry using thermographic phosphor particles. Appl. Phys. B 2013;111(2):155–60.

[154] Heinze J, Meier U, Behrendt T, Willert C, Geigle K-P, Lammel O et al. PLIF Thermometry Based on Measurements of Absolute Concentrations of the OH Radical. Zeitschrift für Physikalische Chemie 2011;225(11-12):1315–41.

[155] Vyrodov AO, Heinze J, Dillmann M, Meier UE, Stricker W. Laser-induced fluorescence thermometry and concentration measurements on NOA: X (0-0) transitions in the exhaust gas of high pressure CH4/air flames. Appl. Phys. B 1995;61(5):409–14.

[156] Giezendanner-Thoben R, Meier U, Meier W, Heinze J, Aigner M. Phase-locked two-line OH planar laser-induced fluorescence thermometry in a pulsating gas turbine model combustor at atmospheric pressure. Appl. Opt. 2005;44(31):6565.

[157] Musazzi S, Perini U (eds.). Laser-Induced Breakdown Spectroscopy. Berlin, Heidelberg: Springer Berlin Heidelberg; 2014.

[158] Lückerath R, Woyde M, Meier W, Stricker W, Schnell U, Magel HC et al. Comparison of coherent anti-Stokes Raman-scattering thermometry with thermocouple measurements and model predictions in both natural-gas and coal-dust flames. Appl. Opt. 1995;34(18):3303–12.

[159] Kapadia BK. Experimental study of oxyfuel combustion for stationary gas turbine applications. Universität Stuttgart; 2017.

[160] Ledesma EB, Kalish MA, Nelson PF, Wornat MJ, Mackie JC. Formation and fate of PAH during the pyrolysis and fuel-rich combustion of coal primary tar. Fuel 2000;79(14):1801–14.

[161] Frigge L, Ströhle J, Epple B. Release of sulfur and chlorine gas species during coal combustion and pyrolysis in an entrained flow reactor. Fuel 2017;201:105–10.

[162] Sepman A, Ögren Y, Gullberg M, Wiinikka H. Development of TDLAS sensor for diagnostics of CO, H2O and soot concentrations in reactor core of pilot-scale gasifier. Appl. Phys. B 2016;122(2):89.

[163] Chrystie RSM, Nasir EF, Farooq A. Propene concentration sensing for combustion gases using quantum-cascade laser absorption near 11 μm. Appl. Phys. B 2015;120(2):317–27.

[164] KC U, Nasir EF, Farooq A. A mid-infrared absorption diagnostic for acetylene detection. Appl. Phys. B 2015;120(2):223–32.

[165] Zhou X, Liu X, Jeffries JB, Hanson RK. Development of a sensor for temperature and water concentration in combustion gases using a single tunable diode laser. Meas. Sci. Technol. 2003;14(8):1459–68.

[166] Schäfer S, Mashni M, Sneider J, Miklós A, Hess P, Pitz H et al. Sensitive detection of methane with a 1.65 μm diode laser by photoacoustic and absorption spectroscopy. Applied Physics B: Lasers and Optics 1998;66(4):511–6.

[167] Zhou X, Jeffries JB, Hanson RK. Development of a fast temperature sensor for combustion gases using a single tunable diode laser. Appl. Phys. B 2005;81(5):711–22.

[168] Pogány A, Klein A, Ebert V. Measurement of water vapor line strengths in the 1.4–2.7μm range by tunable diode laser absorption spectroscopy. Journal of Quantitative Spectroscopy and Radiative Transfer 2015;165:108–22.

[169] Liu X, Jeffries JB, Hanson RK. Measurement of Non-Uniform Temperature Distributions Using Line-of-Sight Absorption Spectroscopy. AIAA Journal 2007;45(2):411–9.

[170] Wang F, Wu Q, Huang Q, Zhang H, Yan J, Cen K. Simultaneous measurement of 2-dimensional H2O concentration and temperature distribution in premixed methane/air flame using TDLAS-based tomography technology. Optics Communications 2015;346:53–63.

[171] Orthwein P. In situ Laserspektroskopie für Vergasungs- und Verbennungsprozesse: Direkte Bestimmung absoluter Gaskonzentrationen und neuer Spektralparameter [Dissertation]. Heidelberg: Ruprecht-Karls-University Heidelberg; 2011.

[172] Schlosser E. Diodenlaser-gestützter In-situ-Nachweis von O 2 und Alkaliatomen zur Optimierung der Hochtemperaturkohlenstaub- und Sondermüllverbrennung sowie der Brandbekämpfung [Dissertation]. Heidelberg: Ruprecht-Karls-University Heidelberg; 2002.

[173] Ebert V. Neue Laserapsorptionsspektroskopische Verfahren zur in-situ Bestimmung von Konzentrationen, Temperaturen und der Verweilzeit [Habilitationsschrift]. Heidelberg: Ruprecht-Karls-University Heidelberg; 2003.

[174] Fernholz T. Entwicklung und Aufbau eines DSP-gestützten Diodenlaserspektrometers zum In-situ-Gasnachweis in industriellen Feuerungsanlagen [Disseration]. Heidelberg: Ruprecht-Karls-University Heidelberg; 2001.

[175] Wang J, Sanders ST, Jeffries JB, Hanson RK. Oxygen measurements at high pressures with vertical cavity surface-emitting lasers. Appl. Phys. B 2001;72(7):865–72.

[176] Ebert V, Fernholz T, Vogel P. Empfindlicher extraktiver Gasnachweis mit Nahinfrarot-Diodenlasern am Beispiel von O_2 und CH_4 (Sensitive Extractive Detection of Oxygen and Methane with Near Infrared Diode Lasers). Technisches Messen 2001;68(10/2001):1527.

[177] Gardiner WC (ed.). Combustion Chemistry. New York, NY: Springer US; 1984.

[178] Wagner S. Ortsaufgelößte in situ Bestimmung absoluter CH_4, H_2O, CO und C_2H_2 Profile in laminaren Gegenstromflammen mittels Diodenlaser-und Cavity-Ring-down Spektroskopy [Dissertation]. Heidelberg: Ruprecht-Karls-University Heidelberg; 2011.

[179] Kitto JB (ed.). Steam: Its generation and use. 41st ed. Barberton, Ohio: Babcock & Wilcox; 2005.

[180] Nicolas J-C, Baranov AN, Cuminal Y, Rouillard Y, Alibert C. Tunable diode laser absorption spectroscopy of carbon monoxide around 235 μm. Appl. Opt. 1998;37(33):7906.

[181] Jainski C. Experimentelle Untersuchung der turbulenten Flamme-Wand-Interaktion [Dissertation]: Optimus Verlag; 2016.

[182] Frigge L, Elserafi G, Ströhle J, Epple B. Sulfur and Chlorine Gas Species Formation during Coal Pyrolysis in Nitrogen and Carbon Dioxide Atmosphere. Energy Fuels 2016;30(9):7713–20.

[183] Wolf C, Stephan A, Fendt S, Spliethoff H. Measuring gaseous HCl emissions during pulverised co-combustion of high shares of straw in an entrained flow reactor. Energy Procedia 2017;120:246–53.

[184] Robinson AL, Junker H, Buckley SG, Sclippa G, Baxter LL. Interactions between coal and biomass when cofiring. Symposium (International) on Combustion 1998;27(1):1351–9.

[185] Nielsen HP, Frandsen FJ, Dam-Johansen K, Baxter LL. The implications of chlorine-associated corrosion on the operation of biomass-fired boilers. Progress in Energy and Combustion Science 2000;26(3):283–98.

[186] Warnatz J. The structure of laminar alkane-, alkene-, and acetylene flames. Symposium (International) on Combustion 1981;18(1):369–84.

[187] Wagner HG. Soot formation in combustion. Symposium (International) on Combustion 1979;17(1):3–19.

[188] Warnatz J, Maas U, Dibble R. Formation of Hydrocarbons and Soot. In: Warnatz J, Maas U, Dibble R, editors. Combustion. Springer Berlin Heidelberg; 2006, p. 277–296.

[189] Ishio H, Minowa J, Nosu K. Review and status of wavelength-division-multiplexing technology and its application. J. Lightwave Technol. 1984;2(4):448–63.

[190] Röpcke J, Mechold L, Käning M, Anders J, Wienhold FG, Nelson D et al. IRMA: A tunable infrared multicomponent acquisition system for plasma diagnostics. Rev. Sci. Instrum. 2000;71(10):3706.

[191] AIAA SciTech (ed.). Common-Path Measurement of H_2O, CO, and CO_2 via TDLAS for Combustion Progress in a Hydrocarbon-Fueled Scramjet. 2016th ed; 2016.

[192] Pogány A, Wagner S, Werhahn O, Ebert V. Development and metrological characterization of a tunable diode laser absorption spectroscopy (TDLAS) spectrometer for simultaneous absolute measurement of carbon dioxide and water vapor. Appl Spectrosc 2015;69(2):257–68.

[193] Tropea C, Foss JF, Yarin AL (eds.). Springer Handbook of Experimental fluid mechanics. Berlin: Springer Science+Business Media; 2007.

[194] Stritzke F, Dreizler A, Schulz C. Absorptionsspektrometrie zur zeitaufgelösten Untersuchung von Ammoniakverteilungen in Abgas [Dissertation]. Darmstadt: Universitäts- und Landesbibliothek Darmstadt; 2017.

[195] Persson P-O, Strang G. Smoothing by Savitzky-Golay and Legendre Filters. In: Arnold DN, Santosa F, Rosenthal J, Gilliam DS, editors. Mathematical Systems Theory in Biology, Communications, Computation, and Finance. New York, NY: Springer New York; 2003, p. 301–315.

[196] Press WH. Numerical recipes in C: The art of scientific computing. 2nd ed. Cambridge: Univ. Press; 2002.

[197] Seidel A, Wagner S, Ebert V. TDLAS-based open-path laser hygrometer using simple reflective foils as scattering targets. Appl. Phys. B 2012;109(3):497–504.

[198] White JU. Long Optical Paths of Large Aperture. J. Opt. Soc. Am. 1942;32(5):285.

[199] Herriott D, Kogelnik H, Kompfner R. Off-Axis Paths in Spherical Mirror Interferometers. Appl. Opt. 1964;3(4):523.

[200] Cai W, Kaminski CF. Tomographic absorption spectroscopy for the study of gas dynamics and reactive flows. Progress in Energy and Combustion Science 2017;59:1–31.

[201] Twynstra MG, Daun KJ. Laser-absorption tomography beam arrangement optimization using resolution matrices. Appl. Opt. 2012;51(29):7059–68.

[202] Rothman LS, Gordon IE, Babikov Y, Barbe A, Chris Benner D, Bernath PF et al. The HITRAN2012 molecular spectroscopic database. Journal of Quantitative Spectroscopy and Radiative Transfer 2013;130:4–50.

[203] Engleman R, Rouse PE. Oscillator strengths from line absorption in a high-temperature furnace— II. The (0,0) band of the $B2\sigma+$—$X2\sigma+$ transition in CN*. Journal of Quantitative Spectroscopy and Radiative Transfer 1975;15(9):831–8.

[204] Casa G, Parretta DA, Castrillo A, Wehr R, Gianfrani L. Highly accurate determinations of CO_2 line strengths using intensity-stabilized diode laser absorption spectrometry. J Chem Phys 2007;127(8):84311.

[205] Toth RA. Measurements of positions, strengths and self-broadened widths of H_2O from 2900 to 8000 cm−1: Line strength analysis of the 2nd triad bands. Journal of Quantitative Spectroscopy and Radiative Transfer 2005;94(1):51–107.

[206] Witzel O, Ebert V. In-situ-Laserabsorptionsspektroskopie zur µs-schnellen Bestimmung von Spezieskonzentrationen und Temperaturen in Verbrennungsmotoren [Dissertation]. Darmstadt: Universitäts- und Landesbibliothek Darmstadt; 2013.

[207] Pogány A, Klein A, Ebert V. Measurement of Water Vapor Line Strengths in the 1.4–2.7 µm Range by Tunable Diode Laser Absorption Spectroscopy. Journal of Quantitative Spectroscopy and Radiative Transfer 2015;165:108–22.

[208] Gharavi M, Buckley SG. Diode laser absorption spectroscopy measurement of linestrengths and pressure broadening coefficients of the methane $2\nu3$ band at elevated temperatures. J Mol Spectrosc 2005;229(1):78–88.

[209] El Hachtouki R, Auwera JV. Absolute Line Intensities in Acetylene: The 1.5-µm Region. J Mol Spectrosc 2002;216(2):355–62.

[210] Malathy Devi V, Chris Benner D, Smith MAH, Mantz AW, Sung K, Brown LR et al. Spectral line parameters including temperature dependences of self- and air-broadening in the $2\leftarrow0$ band of CO at 2.3µm. Journal of Quantitative Spectroscopy and Radiative Transfer 2012;113(11):1013–33.

[211] Hashemi R, Predoi-Cross A, Dudaryonok AS, Lavrentieva NN, Vandaele AC, Vander Auwera J. CO 2 pressure broadening and shift coefficients for the 2–0 band of 12 C 16 O. J Mol Spectrosc 2016;326:60–72.

[212] Rosenmann L, Hartmann JM, Perrin MY, Taine J. Accurate calculated tabulations of IR and Raman CO_2 line broadening by CO_2, H_2O, N_2, O_2 in the 300-2400-K temperature range. Appl Opt 1988;27(18):3902–6.

[213] Durry G, Zeninari V, Parvitte B, Le barbu T, Lefevre F, Ovarlez J et al. Pressure-broadening coefficients and line strengths of H_2O near 1.39μm: Application to the in situ sensing of the middle atmosphere with balloonborne diode lasers. Journal of Quantitative Spectroscopy and Radiative Transfer 2005;94(3-4):387–403.

[214] Moretti, Sasso, Gianfrani, Ciurylo. Collisional-Broadened and Dicke-Narrowed Lineshapes of H(2)(16)O and H(2)(18)O Transitions at 1.39 μm. J Mol Spectrosc 2001;205(1):20–7.

[215] Hunsmann, S., Wagner, S., Saathoff, H., Moehler, O., Schurath, U., Ebert, V., and Möhler, O. Messung der Linienstärken und Druckverbreiterungskoeffizienten von H_2O-Absorptionslinien im 1.4 μm - Band. VDI BERICHTE;2006(1959):149–64.

[216] Ptashnik IV, Smith KM, Shine KP. Self-broadened line parameters for water vapour in the spectral region 5000–5600cm−1. J Mol Spe□tros□2005;232(2):186–201.

[217] Brown, Plymate. Experimental Line Parameters of the Oxygen A Band at 760 nm. J Mol Spectrosc 2000;199(2):166–79.

[218] Korb CL, Hunt RH, Plyler EK. Measurement of Line Strengths at Low Pressures—Application to the 2–0 Band of CO. J Chem Phys 1968;48(9):4252–60.

[219] Bouanich J-P, Brodbeck C. Mesure des largeurs et des deplacements des raies de la bande 0 → 2 de CO autoperturbe et perturbe par N_2, O_2, H_2, HCl, NO et CO_2. Journal of Quantitative Spectroscopy and Radiative Transfer 1973;13(1):1–7.

[220] Predoi-Cross A, Bouanich JP, Benner DC, May AD, Drummond JR. Broadening, shifting, and line asymmetries in the 2←0 band of CO and CO–N_2: Experimental results and theoretical calculations. J Chem Phys 2000;113(1):158–68.

[221] Malathy Devi V, Predoi-Cross A, Chris Benner D, Smith MAH, Rinsland CP, Mantz AW. Self- and H2-broadened width and shift coefficients in the 2←0 band of 12C16O: Revisited. J Mol Spectrosc 2004;228(2):580–92.

[222] Sung K, Varanasi P. -broadened half-widths and -induced line shifts of relevant to the atmospheric spectra of Venus and Mars. Journal of Quantitative Spectroscopy and Radiative Transfer 2005;91(3):319–32.

[223] Arteaga SW, Bejger CM, Gerecke JL, Hardwick JL, Martin ZT, Mayo J et al. Line broadening and shift coefficients of acetylene at 1550nm. J Mol Spectrosc 2007;243(2):253–66.

[224] Minutolo P, Corsi C, D'Amato F, Rosa M de. Self- and foreign-broadening and shift coefficients for C2H2 lines at 1.54 μm. The European Physical Journal D 2001;17(2):175–9.

[225] Li G, Asfin RE, Domanskaya AV, Ebert V. He-broadening and shifting coefficients of HCl lines in the (1←0) and (2←0) infrared transitions. Mole□ular Physi□s 2018;35:1–8.

[226] Li G, Serdyukov A, Gisi M, Werhahn O, Ebert V. FTIR-based measurements of self-broadening and self-shift coefficients as well as line strength in the first overtone band of HCl at 1.76μM. Journal of Quantitative Spectroscopy and Radiative Transfer 2015;165:76–87.

[227] Walcher W, Elbel M. Praktikum der Physik: Mit 88 Versuchen, 14 Tabellen im Text, einem Tabellenanhang und einem ausklappbaren Periodensystem der Elemente. 9th ed. Wiesbaden: Vieweg + Teubner; 2009.

[228] Régalia-Jarlot L, Zéninari V, Parvitte B, Grossel A, Thomas X, Heyden P von der et al. A complete study of the line intensities of four bands of CO2 around 1.6 and 2.0μm: A comparison between Fourier transform and diode laser measurements. Journal of Quantitative Spectroscopy and Radiative Transfer 2006;101(2):325–38.

[229] Toth RA, Miller CE, Malathy Devi V, Benner DC, Brown LR. Air-broadened halfwidth and pressure shift coefficients of 12C16O2 bands: 4750–7000cm−1. J Mol Spectrosc 2007;246(2):133–57.

[230] Toth RA. Extensive measurements of H216O line frequencies and strengths: 5750 to 7965 cm(-1). Appl Opt 1994;33(21):4851–67.

[231] Witzel O, Ebert V. In-situ-Laserabsorptionsspektroskopie zur µs-schnellen Bestimmung von Spezieskonzentrationen und Temperaturen in Verbrennungsmotoren. Darmstadt: Universitäts- und Landesbibliothek Darmstadt; 2013.

[232] Varanasi P. Collision-broadened half-widths and shapes of methane lines. Journal of Quantitative Spectroscopy and Radiative Transfer 1971;11(11):1711–24.

[233] Dufour G, Hurtmans D, Henry A, Valentin A, Lepère M. Line profile study from diode laser spectroscopy in the 12CH4 2v3 band perturbed by N2, O2, Ar, and He. J Mol Spectrosc 2003;221(1):80–92.

[234] Geyer D, Kempf A, Dreizler A, Janicka J. Scalar dissipation rates in isothermal and reactive turbulent opposed-jets: 1-D-Raman/Rayleigh experiments supported by LES. Proceedings of the Combustion Institute 2005;30(1):681–9.

[235] Chen S, Zheng C. Counterflow diffusion flame of hydrogen-enriched biogas under MILD oxy-fuel condition. International Journal of Hydrogen Energy 2011;36(23):15403–13.

[236] Cheng Z, Wehrmeyer JA, Pitz RW. Experimental and numerical studies of opposed jet oxygen-enhanced methane diffusion flames. Combustion Science and Technology 2006;178(12):2145–63.

[237] Naik SV, Laurendeau NM. Quantitative laser-saturated fluorescence measurements of nitric oxide in counter-flow diffusion flames under sooting oxy-fuel conditions. Combustion and Flame 2002;129(1-2):112–9.

[238] Liu CY, Chen G, Sipöcz N, Assadi M, Bai XS. Characteristics of oxy-fuel combustion in gas turbines. Applied Energy 2012;89(1):387–94.

[239] Baroncelli M, Felsmann D, Kruse S, Beeckmann J, Pitsch H. Extinction Strain Rates for Coal Light Volatiles Under Oxy-Fuel Conditions. Proceedings of the China National Symposium on Combustion 2016;2016.

[240] Baroncelli M, Felsmann D, Hansen N, Pitsch H. Investigating the Effect of Oxy-Fuel combustion and Light coal Volatiles Interaction: A Mass Spectrometric Study. Fuel;2018:submitted.

[241] Wolfhard HG, Parker WG. A Spectroscopic Investigation into the Structure of Diffusion Flames. Proc. Phys. Soc. A 1952;65(1):2–19.

[242] Smyth KC, Miller JH, Dorfman RC, Mallard WG, Santoro RJ. Soot inception in a methane/air diffusion flame as characterized by detailed species profiles. Combustion and Flame 1985;62(2):157–81.

[243] Smyth KC, Tjossem PJH, Hamins A, Miller JH. Concentration measurements of OH and equilibrium analysis in a laminar methane-air diffusion flame. Combustion and Flame 1990;79(3-4):366–80.

[244] Hamins A, Anderson DT, Houston MJ. Mechanistic Studies of Toluene Destruction in Diffusion Flames. Combust. Sci. Technol. 1990;71(4-6):175–95.

[245] Norton TS, Smyth KC, Miller JH, Smooke MD. Comparison of Experimental and Computed Species Concentration and Temperature Profiles in Laminar, Two-Dimensional Methane/Air Diffusion Flames. Combust. Sci. Technol. 1993;90(1-4):1–34.

[246] Houston Miller J, Elreedy S, Ahvazi B, Woldu F, Hassanzadeh P. Tunable diode-laser measurement of carbon monoxide concentration and temperature in a laminar methane-air diffusion flame. Appl. Opt. 1993;32(30):6082–9.

[247] Buchholz B, Ebert V. Offsets in fiber-coupled diode laser hygrometers caused by parasitic absorption effects and their prevention. Meas. Sci. Technol. 2014;25(7):75501.

[248] Peters N. Numerical and asymptotic analysis of systematically reduced reaction schemes for hydrocarbon flames. In: Glowinski R, Larrouturou B, Temam R, editors. Numerical Simulation of Combustion Phenomena. Berlin/Heidelberg: Springer-Verlag; 1985, p. 90–109.

[249] Bass CA, Barat RB, Lemieux PM. Identification of an ideal reactor model in a secondary combustion chamber. AIChE J. 2003;49(10):2619–30.

[250] Heil P, Toporov D, Stadler H, Tschunko S, Förster M, Kneer R. Development of an oxycoal swirl burner operating at low O_2 concentrations. Fuel 2009;88(7):1269–74.

[251] Bendat JS, Piersol AG. Random data: Analysis and measurement procedures. Hoboken, New Jersey: Wiley a John Wiley & Sons Inc; 2010.

[252] Fox RF, Gatland IR, Roy R, Vemuri G. Fast, accurate algorithm for numerical simulation of exponentially correlated colored noise. Phys. Rev. A 1988;38(11):5938–40.

[253] Goodwin DG, Moffat HK, Speth RL. Cantera: An Object-Oriented Software Toolkit For Chemical Kinetics, Thermodynamics, And Transport Processes. Version 2.3.0. Zenodo; 2017.

[254] Somers LMT. The simulation of flat flames with detailed and reduced chemical models. Technische Universiteit Eindhoven; 1994.

[255] Becker LG, Langenthal T von, Pielsticker S, Böhm B, Kneer R, Dreizler A. Experimental investigation of particle-laden flows in an oxy-coal combustion chamber for non-reacting conditions. Fuel 2019;235:753–62.

[256] Fisher EM, Dupont C, Darvell LI, Commandré J-M, Saddawi A, Jones JM et al. Combustion and gasification characteristics of chars from raw and torrefied biomass. Bioresour Technol 2012;119:157–65.

[257] van der Stelt MJC, Gerhauser H, Kiel JHA, Ptasinski KJ. Biomass upgrading by torrefaction for the production of biofuels: A review. Biomass and Bioenergy 2011.

[258] Ohliger A. Experimentelle und theoretische Untersuchung der Torrefizierung von Buchenholz [Dissertation]. Aachen: RWTH Aachen; 2014.

[259] Prins M, Ptasinski K, Janssen F. From coal to biomass gasification: Comparison of thermodynamic efficiency. Energy 2007;32(7):1248–59.

[260] van Krevelen DW. Coal: Typology, physics, chemistry, constitution. 3rd ed. Amsterdam: Elsevier; 1993.

[261] Heuer S, Senneca O, Wütscher A, Düdder H, Schiemann M, Muhler M et al. Effects of oxy-fuel conditions on the products of pyrolysis in a drop tube reactor. Fuel Processing Technology 2016;150:41–9.

[262] Toporov D, Bocian P, Heil P, Kellermann A, Stadler H, Tschunko S et al. Detailed investigation of a pulverized fuel swirl flame in CO_2/O_2 atmosphere. Combustion and Flame 2008;155(4):605–18.

[263] Zabrodiec D, Hees J, Massmeyer A, Vom Lehn F, Habermehl M, Hatzfeld O et al. Experimental investigation of pulverized coal flames in CO 2 /O 2 - and N 2 /O 2 -atmospheres: Comparison of solid particle radiative characteristics. Fuel 2017;201:136–47.

[264] Smart JP, Patel R, Riley GS. Oxy-fuel combustion of coal and biomass, the effect on radiative and convective heat transfer and burnout. Combustion and Flame 2010;157(12):2230–40.

[265] Skeen SA, Kumfer BM, Axelbaum RL. Nitric Oxide Emissions during Coal and Coal/Biomass Combustion under Air-Fired and Oxy-fuel Conditions. Energy Fuels 2010;24(8):4144–52.

[266] Möller G, Maßmeyer A, Hees J, Zabrodiec D, Habermehl M, Kneer R. Chemiluminescence Mcasurements in a Pulverized Fuel Furnace of Pure Torrefied Biomasses in Comparison to Solid Fossil Fuels. In: 42nd International Technical Conference on Clean Energy, 2017, Clearwater, Florida, USA.

[267] Linstrom PJ, Mallard WG (eds.). NIST Chemistry WebBook: NIST Standard Reference Database Number 69.

[268] Hees J, Zabrodiec D, Massmeyer A, Hatzfeld O, Kneer R. Experimental Investigation into the Influence of the Oxygen Concentration on a Pulverized Coal Swirl Flame in Oxy-fuel Atmosphere. Fuel Special Issue 2018:submitted.

[269] Hees J, Zabrodiec D, Maßmeyer A, Hatzfeld O, Habermehl M, Kneer R. Kohlenstaubverbrennung in Oxyfuel-Atmosphäre - Einfluss der Oxidatorzusammensetzung auf den Verbrennungsprozess. In: 28. Deutscher Flammentag, 2017, Darmstadt, Germany.

Own Publications

[E1] **Bürkle** S., Becker, L.G., Agizza, M.A., Dreizler. A., Ebert, V., Wagner, S. In-situ measurement of residence time distributions in a turbulent oxy-fuel gas-flame combustor. Exp Fluids 2017;58(7):647.

[E2] **Bürkle** S., Walter, N., Wagner S. Laser-based measurements of pressure broadening and pressure shift coefficients of combustion-relevant absorption lines in the near-infrared region. Appl Phys B 2018;124(6):121.

[E3] **Bürkle** S., Greifenstein, M., Wagner, S., Dreizler, A., Ebert, V. Optical Sensing of Turbine Inlet Temperature in a Pressurized Gas Turbine Combustor. Laser Applications to Chemical, Security and Environmental Analysis, 2016.

[E4] **Bürkle,** S., Biondo, L., Ding, C.-P., Honza, R., Ebert, V., Böhm, B., Wagner, S. In-Cylinder Temperature Measurements in a Motored IC Engine using TDLAS. Flow Turbulence Combust 2018;101(1):139–59.

[E5] **Bürkle,** S., Dreizler, A., Ebert, V., Wagner, S. Experimental Comparison of a 2D Laminar Diffusion Flame under Oxy-Fuel and Air Atmosphere. Fuel; 2018 (212):302-308.

[E6] **Bürkle,** S., Becker, L.G., Dreizler, A., Wagner, S. Experimental investigation of the flue gas thermochemical composition of an oxy-fuel swirl burner. Fuel; 2018(231):61-72.

[E7] Doost, A.S., Ries, F., Becker, L.G., **Bürkle,** S., Wagner, S., Ebert, V., Dreizler, A., Di Mare, F., Sadiki, A., Janicka, J. Residence time calculations for complex swirling flow in a combustion chamber using large-eddy simulations. Chemical Engineering Science 2016;156:97–114.

[E8] **Bürkle,** S., Becker, L., Dreizler, A., Wagner, S. An improved TDLAS technique to measure residence time distributions in particle loaded combustion chambers. Laser Applications to Chemical, Security and Environmental Analysis, 2018.

[E9] Agizza, M.A., **Bürkle,** S., Becker, L.G., Greifenstein, M., Bagheri, G., Doost, S., Faravelli, T., Janicka, J., Wagner, S., Dreizler, A. Chemical Reactors Network Modelling in Close to Reality Combustion Systems. In: 10th Mediterranian Combustion Symposium, 2017.

[E10] **Bürkle,** S., Becker, Agizza, M.A., Dreizler, A. Wagner, S., A Comparison of Two Measurement Strategies to obtain the Residence Time Distribution in Combustion Chambers. Applied Optics, **accepted** 2019.

[E11] **Bürkle,** S., Möller, G., Hees, J., Zabrodiec, D., Maßmeyer, A., Rauen, D., Ebert, V., Kneer, R., Dreizler, A., Wagner, S. Laserbasierte Messung von Temperatur und Spezieskonzentrationen bei der Verbrennung von torrefizierter Biomasse in einer 100 kW_{th}-Staubfeuerungsversuchsanlage, VDI Flammentag;2017.

Additional Conference Contributions

Oral Presentation
Laser-Based Measurements of Temperature and Species concentrations in a 20 kW Oxyfuel burner. Bürkle, S., Becker, L.G., Dreizler, A., Wagner, S. Fifth International Education Forum on Environment and Energy Science, San Diego, USA, December 2016.

A complementary approach for the experimental investigation of solid oxy-fuel combustion using optical diagnostics. Bürkle, S., Hees, J. Second International Oxyflame Workshop, Bochum, Germany, February 2018

Visual Presentation
TDLAS-Based In-Situ Measurement of Gas Residence Time Distributions in an Oxyfuel Gas-Flame Assisted Coal Particle Combustor. Bürkle, S., Becker, L.G., Dreizler, A., Ebert, V., Wagner, S. Field Laser Applications in Industry and Research, Aix-les-Bains, France, September 2016.

Impact of Temperature Variations in flames on TDLAS Two-Line Thermometry studied with Rayleigh-Scattering-Spectroscopy. Bürkle, S., Köser, J., Böhm, B., Dreizler, A., Ebert, V., Wagner, S. Field Laser Applications in Industry and Research, Firenze, Italy, May 2014

Curriculum Vitae

Sebastian Bürkle, M.Sc. M.Sc.

Born on March, 5th 1987 in Erbach/Odenwald
Alter Darmstädter Weg 35, 64380 Rossdorf, Germany
Email: sebastian.buerkle@gmail.com
Tel: +49 151 23670477

Employment

2015 – present	*TU Darmstadt* **Managing Director** Collaborative Research Cluster/Transregio 150
2013 – present	*TU Darmstadt* **Research Assistant** *Projects:* AG-Turbo (BMWI) Laser-based temperature measurements in gas turbines Partners: Alstom/GE, Rolls Royce Germany, DLR Collaborative Research Cluster/Transregio 129 (DFG) Laser-based investigation of coal combustion within oxy-fuel atmosphere Partners: RWTH Aachen, Ruhr-Universität Bochum Collaborative Research Cluster/Transregio 150 (DFG) Measurement techniques for near-wall reactive flows Partners: Karlsruhe Institute of Technology

Education and qualifications

2013 – present	*TU Darmstadt* **Doctoral candidate Mechanical Engineering**
2010 – 2014	*TU Darmstadt* **M.Sc. Mechanical and Process Engineering** Major field of study: Aerodynamics and turbines Minor field of study: Economics (final grade 1.09) **M.Sc. Physics** Major field of studies: High energy density matter (final grade 1.17)
2010 – 2011	*University of California, Berkeley*, USA Two semesters abroad Degree course Mechanical Engineering

2006 – 2010	*TU Darmstadt* **B.Sc. Mechanical and Process Engineering** (final grade 1.9) **B.Sc. Physics** (final grade 1.9)
1999 – 2006	*Gymnasium Michelstadt* Abitur (final grade 1.7, best graduate in physics)

Internships

2012	**Alstom S.A.** in Birr, Switzerland
2008	**German Aerospace Center** in Lampoldshausen
2006	**Mühlhäuser GmbH & Co. Kg** in Michelstadt

Additional skills and achievements

<u>Scholarships</u>

2008 – 2012	**Studienstiftung des deutschen Volkes**
2010 – 2011	**DAAD – ISAP Scholarship** for a year abroad at the University of California, Berkeley, USA

<u>Awards</u>

2013	**Advancement – Award by the Odenwald-Academy** for Master-Theses
2006	**Prof. Walter Masing – Award** by Lions Club Odenwald
2006	Award for **best high school graduate** in Physics
2004	**Youth and future – Award** by Rotary Club Odenwald